Number Theory and Its Applications

Number Theory and Its Applications

Fuhuo Li
Sanmenxia SuDa Transportation Energy Saving Technology Co., Ltd., China

Nianliang Wang
Shangluo University, China

Shigeru Kanemitsu
Kinki University, Japan

 **World Scientific**

NEW JERSEY · LONDON · SINGAPORE · BEIJING · SHANGHAI · HONG KONG · TAIPEI · CHENNAI

Published by

World Scientific Publishing Co. Pte. Ltd.
5 Toh Tuck Link, Singapore 596224
USA office: 27 Warren Street, Suite 401-402, Hackensack, NJ 07601
UK office: 57 Shelton Street, Covent Garden, London WC2H 9HE

British Library Cataloguing-in-Publication Data
A catalogue record for this book is available from the British Library.

NUMBER THEORY AND ITS APPLICATIONS

ISBN 978-981-4425-63-6

Printed in Singapore.

To Professor Andrzej Schinzel on his seventy-fifth birthday
with deep respect

Preface

Number theory is an old subject in the sense that it started as one of the most intelligent subjects in ancient Greek disciplines and yet is a new subject whereby it is ever-developing and continues to attract the interest of the most intelligent audience. Owing to this, at any time in history, there is an enormous amount of results assembled and included in textbooks. Therefore, there are a few different presentations of the theory.

The titles of chapters shown herein will already give a feeling of this book. We try to present as many viewpoints as possible of an important phenomenon and we emphasize the role of symmetry—functional equation of the associated zeta-function—throughout the book. We also try to present as many worked-out examples and exercises as possible so as to enhance the reader's comprehension.

The following illustrates the feature of the book: *starting from the very basics and warp into the space of new interest; from the ground state to the excited state.*

We treat the Euler function in several different places: as the number of generators of a finite cyclic group, as one counting the order of the multiplicative group of reduced residue classes modulo q, as the order of the Galois group of the cyclotomic field, and as a result, the degree of the cyclotomic field.

We give several independent proofs of the Gauss' quadratic reciprocity law, thereby stressing the way the functional equation is being applied. We shall give four different proofs in all. Gauss himself gave seven different proofs. In Chapter 1, we give a proof of F. Keune using the theory of finite fields. In Chapter 4, a proof given by Auslander, Tolimieri and Winograd which rests on the representation theory of finite nilpotent groups and can work for the proof of Hecke's reciprocity law for algebraic number fields. In Chapter 5, we give two independent proofs—one depending on the Zolotareff-Frobenius theorem and the reciprocity law for Dedekind sums, and the other on the theta-transformation formula. As can be seen in the theorem in Chapter 6, the theta-transformation formula is equivalent to the functional equation for the relevant zeta-function (and the reciprocity law for the Dedekind sums is also a consequence of the functional equation), we can see the remarkable catalytic effect of the zeta-symmetry.

We establish a few guiding principles which are clear to experts but have not been formulated explicitly before: the gcd (greatest common divisor) principle, exhaustion principle, relative primality principle, visualization principle, etc. We give several different applications of the relative primality principle as well as the Möbius inversion formula which is useful in the context of physics too.

We also provide the reader with some basic knowledge from other disciplines including functional analysis and representation theory (Chapters 3 and 4) and some basics in control systems (Chapter 6). Since they are linear systems, the vector space structure is to be incorporated and it is well-versed in Chapters 1 and 2.

Somewhat novel is the presentation of the concatenation rule (cascade connection in the case of control theory). The compositum of actions of the special linear group in Exercise 32 may be thought of as a concatenation and so is the polymerization of DNA's viewed from the point of view of formal language theory. In Chapter 6, we present the theory of chain scattering representation of a plant which renders the cascade connection much easily accessible and leads to the action of the symplectic group.

Most of the chapters can be read independently of others. We intentionally repeat the same formulas or expressions if they are very essential. For example, the definition of the Riemann zeta-function appears in several places although the proper one is given by (5.1). So is the Dirichlet L-function, the proper definition being given by (5.3) in Chapter 5. In other places, where the L-function is referred to, one is supposed to view it as kin to the Riemann zeta-functions, with Euler product and Dirichelt series expansion. Also Dirichlet characters are redundantly given in Chapters 1, 3, 4 and 5.

Chapter 1 gives an introduction to algebra accessible to undergraduates and could be used as a material for a semester. It can be skipped if the reader is equipped with basic knowledge in algebra. Some examples and exercises may be of some interest in their own right, though.

Chapter 2 gives only a brief introduction to (global) algebraic number theory and can be skipped by experienced readers. No local theory is presented and so, the interested reader may consult more advanced books on the subject.

Parts of Chapters 1 and 2 are expanded version of Appendices to [Vis2], which can be read beneficially together.

Chapter 3 provides an elementary introduction to analytic number theory, with emphasis on the use of Stieltjes integrals. There we shall also give three types of generating functions for arithmetic functions, power series, Dirichlet series and Euler products. Arithmetic functions are treated rather algebraically. Another special feature is the presentation of part of Hilbert space and number theory a là N. P. Romanoff and A. Wintner. This chapter is also rather sketchy and the interested reader may read more advanced books.

Chapter 4 is devoted to the presentation of the theory of dual group of a finite Abelian group with additional materials and could be a rough guide for representation theory. We try to present it in finite terms.

Chapter 5 occupies a rather special position, giving some results on the Dirichlet L-functions with both primitive and imprimitive characters. Dirichlet L-functions are intermediate objects lying between the most basic Riemann zeta-function and more sophisticated zeta- and L-functions and are bridges between them. They are the keys to the proof of Dirichlet's prime number theorem, whose proof contains the important Dirichlet class number formula. The values of Dirichlet characters lying in cyclotomic fields which are the starters for class field theory. The best analytic treatment of them is given in [Prac]. This chapter treats somewhat more advanced material and is meant for researchers.

Finally, Chapter 6 gives some accessible expounding of the theory of control systems, which have never been stated in number theory books, but as one can see, it can be number-theorized and will give rise to a nice interactive cooperation, as with the theory of Auslander, Tolimieri and Winograd which treats the intimate interaction between Hecke's reciprocity law—number theory (pure mathematics) and a multi-dimensional Cooley-Tukey algorithm—digital signal processing (engineering). There, emphasis is placed on the nest structure of many notions; unity-feed back system $\subset$ H^∞-control system. There, by resorting to the notion of the power function, one is referred to the association of the mean values of the zeta-function and promises a fertile land lying ahead of us. The colossal theorem in Chapter 6 regarding the modular relation is one from the forthcoming [MRSK].

We remark a funny coincidence of the abuse of language, PID, which is an abbreviation of Principal Ideal Domain on one hand, and Proportional, Integral Differential control on the other.

The authors would like to thank Ms Ji Zhang and Ms Lai Fun Kwong who have helped them through the process with their efficient editorial skills.

The third author would like to express his hearty thanks to Professor Michel Waldschmidt for his constant encouragement and stimulating discussions. The third author also would like to thank his close friend Professor H^2 Chan for his enlightening remarks and the passion for research by which he got infected.

The Authors
2012

Contents

Chapter 1

Elements of algebra

In this chapter we make a brief description of modern algebra to such an extent that is necessary to make algebraic elucidation of some important materials in elementary number theory which are usually treated in an intuitive and elementary way. Algebraic preliminaries can be found in many textbooks (cf. e.g. [Hatt]) but our presentation gives a quick introduction through groups. We recall elements of algebra, groups, rings (Euclidean domains) and fields including the notion of orbits. These are then applied to deducing that the ring of rational integers is a UFD (Euclidean → PID → UFD), a generalization of Fermat's little theorem, which then gives rise to Fermat's little theorem, etc. These abstract notions are also used in Chapter 2 in the case of an algebraic number field, thus generalizing elementary number theory to algebraic number theory. This chapter can serve as a mini-course on abstract algebra for undergraduates.

1.1 Preliminaries

The notion of equivalence classes is one of the most fundamental in modern mathematics and in daily life matters as well.

Definition 1.1. A (binary) relation $\sim$ defined on a set $X \neq \varnothing$ is called an **equivalence relation** if it satisfies
(i) (reflexive law) $x \sim x$.
(ii) (symmetric law) If $x \sim y$, then $y \sim x$.
(iii) (transitive law) If $x \sim y$ and $y \sim z$, then $x \sim z$.
If $x \sim y$, then x is said to be equivalent to y or conversely in view of (ii). The set C_x of all elements y equivalent to x is called an **equivalence class** containing x:

$$C_x = \{y \in X \mid y \sim x\}. \tag{1.1}$$

The following theorem provides us with a **classification** $X/\sim$ of the set X into disjoint union of mutually inequivalent classes.

Theorem 1.1. *The following statements are equivalent.*

(i) $x \sim y.$

(ii) $C_x = C_y.$

(iii) $C_x \cap C_y \neq \varnothing.$

Let Λ be a **complete set of representatives** of X with respect to the equivalence relation $\sim$ (that is, the set of all inequivalent elements of X w.r.t. $\sim$). Then $X/\sim = \cup_{\lambda \in \Lambda} C_\lambda$ is a classification of X.

This notion of equivalence classes is one of the most important notions in the whole of science and many concrete examples will be given in what follows. Here we state a few most familiar examples.

- Two vectors (directed line segments in the plane or in the space) are equivalent if they can be overlapped via translation. A complete set of representatives may be given as the set of all directed line segments with initial points at the origin, which may be denoted by the coordinates of the terminating point.

- Two planar triangles are equivalent (congruent) if they can be overlapped via translation, rotation and reflection (the Euclidean group). Thus we may consider only one triangle as one representing the class of all congruent triangles.

- Two integers a, b are said to be congruent modulo a given natural number m written: $a \equiv b \bmod m$ if there is an integer q such that $a - b = mq$. This example will repeatedly appear.

- Isomorphic images constitute equivalence classes. We often identify an algebraic sub-system with its isomorphic image in a bigger system and call it an **embedding**.

On these examples light may be shed from a more advanced standpoint, i.e. as the group action. In the second example, the group acting is the **Euclidean group** and in the first, it is the subgroup of translations. In the third it is the subgroup consisting of all multiples of m (cf. Example 1.5).

We denote the set of natural numbers by $\mathbb{N}$:

$$\mathbb{N} = \{1, 2, 3, \cdots\}$$

in which we assume the principle of mathematical induction (complete induction) and the gcd principle hold.

Proposition 1.1. *Suppose X is a subset of $\mathbb{N}$. Then if (i) $1 \in X$ and (ii) if $1, \cdots, n-1 \in X$, then $n \in X$ holds, then $X = \mathbb{N}$.*

This implies the important principle.

Theorem 1.2. *Suppose X is a non-empty subset of $\mathbb{N}$. Then X has the minimal element $\min X$.*

Proof. Suppose there does not exist $\min X$. Then $1 \notin X$. If $1, \cdots, n-1 \notin X$, then $n \notin X$ and mathematical induction implies that $X = \varnothing$, a contradiction. $\square$

Also we have the **Euclidean division**.

Theorem 1.3. *Let n be a natural number. For a natural number m, there exist natural numbers q and r such that*

$$m = qn + r, \quad r \leq n. \tag{1.2}$$

Proof. We have only to note that

$$\mathbb{N} = \{1, \cdots, n\} \cup_{q=1}^{\infty} \{qn+1, \cdots, qn+n\}. \quad \square$$

Theorem 1.4. *We have the general **gcd principle**:*

Let $P(n)$ denote a proposition with respect to $n \in \mathbb{N}$ having the property that if $P(m)$ and $P(n)$ are true, then so is $P((m,n))$, (m,n) designating the g.c.d. of m and n. Then the least f for which $P(f)$ is true must divide other n for which $P(n)$ is true (i.e. $n = fq$ for some $q \in \mathbb{N}$).

Proof. Let $X = \{n \in \mathbb{N} \mid P(n) \text{ true}\}$. We may suppose that $X \neq \varnothing$. Then $\min X := f$ exists. For any $n \in X$, we may write

$$n = fq + r, \quad r, q \in \mathbb{N}, \quad r \leq f.$$

By the property mentioned above, $P((f,r))$ must be true because $P((n,f))$ is true. If $r < f$ then $(f,r) < f$, a contradiction. Hence we must have $r = f$, and $n = f \cdot (q+1)$. $\square$

The set of **rational integers** is denoted by $\mathbb{Z}$:

$$\mathbb{Z} = \{0, \pm 1, \pm 2, \pm 3, \cdots \},$$

which we shall construct from $\mathbb{N}$ as equivalence classes.

Definition 1.2. Let $X_1 \times X_2$ denote the **Cartesian product** (or **direct product**) of two sets X_1 and X_2, i.e. the set of all **ordered pairs** (x_1, x_2) with $x_i \in X_i$, or what we call coordinates. We may consider an infinite Cartesian product $\prod_{\lambda \in \Lambda} X_\lambda$, where X_λ are arbitrary sets and Λ is an index set. An element a of $\prod_{\lambda \in \Lambda} X_\lambda$ is a mapping which maps an index λ to the λ-th component $a_\lambda \in X_\lambda$ and denoted $a = (a_\lambda)$.

We introduce an equivalence relation in $\mathbb{N} \times \mathbb{N}$: two points (m, n) and (m', n') are said to be equivalent if $m + n' = m' + n$.

In the following passages we anticipate Definition 1.5 of groups and subgroups.

Exercise 1. Prove that this is an equivalence relation and describe equivalence classes. Also, defining addition by $C_{m_1, n_1} + C_{m_2, n_2} = C_{m_1 + m_2, n_1 + n_2}$, prove that $\mathbb{Z}$ forms an Abelian group in the sense of Definition 1.5.

Solution. It is immediate to check the equivalence conditions and group conditions and therefore are left to the reader. An equivalence class is of one of three types $C_{m+n, n}, C_{m, m}, C_{n, m+n}$ and these are regarded as $m, 0, -m$. Each class depends only on m because e.g. $C_{m+n, n} = C_{m+n', n'}$.

Exercise 2. In the set $\mathbb{Z}$ of equivalence classes, we introduce multiplication according to the signs:

$$C_{m_1+n_1, n_1} C_{m_2+n_2, n_2} = C_{m_1 m_2 + n_3, n_3} \leftrightarrow m_1 \cdot m_2, \quad m_1 > 0, m_2 > 0;$$
$$C_{n_1, m_1+n_1} C_{n_2, m_2+n_2} = C_{m_1 m_2 + n_3, n_3} \leftrightarrow m_1 \cdot m_2, \quad m_1 < 0, m_2 < 0;$$
$$C_{m_1+n_1, n_1} C_{n_2, m_2+n_2} = C_{n_3, m_1 m_2 + n_3} \leftrightarrow m_1 \cdot m_2, \quad m_1 > 0, m_2 < 0;$$
$$C_{m_1+n_1, n_1} C_{n_2, n_2} = C_{n_3, n_3} \leftrightarrow m_1 \cdot 0 = 0,$$

i.e. $C_{m,m}$ acts as 0. Prove that $\mathbb{Z}$ forms a commutative semi-group and that they satisfy the distributive law: $a(b + c) = ab + ac$.

Definition 1.3. If a non-empty set R with addition and multiplication is an Abelian group with respect to addition with the identity 0 and a semi-group with respect to multiplication and satisfies the distributive laws $a(b + c) = ab + ac, (b + c)a = ba + ca$, it is called a **ring**. A ring is called an **integral**

domain if there are no elements $a \neq 0, b \neq 0$ such that $ab = 0$ (zero-divisors). R is called a commutative ring if multiplication is commutative. We consider only **commutative rings** with the multiplicative unit 1. An additive subgroup I of R is called an **ideal** if it is closed under the action of R, i.e. if $r \in R$ and $a \in I$ imply $ra \in R$. The division is not always possible in a ring. If an element $a \in R$ has its multiplicative inverse a^{-1}, then it is called an invertible element. The set of all invertible elements in R is denoted by $R^{\times}$ (or also R^*), called the **unit group**. If $R^{\times} = R - \{0\}$, then R is called a **field**.

Familiar examples of fields are $\mathbb{Q}$, $\mathbb{R}$, $\mathbb{C}$, the field of rational numbers, the field of real numbers and the field of complex numbers. $\mathbb{Q}$ will be introduced in Exercise 3 and $\mathbb{C}$ in Exercise 7, while the real number field $\mathbb{R}$ will be introduced in Chapter 2.

The next theorem implies that the ring of integers is a **principal ideal domain** (abbreviated as PID) to the effect that all ideals are generated by one element, such an ideal being called a principal ideal, thus the name.

Theorem 1.5. *Let* $\mathfrak{i}$ *be an additive subgroup of* $\mathbb{Z}$, *i.e.* $a + b \in \mathfrak{i}, a - b \in \mathfrak{i}$ *for* $a, b \in \mathfrak{i}$. *Then there is a non-negative integer* d *such that*

$$\mathfrak{i} = \{kd \mid k \in \mathbb{Z}\} = (d). \tag{1.3}$$

Proof. Take an element $a \in \mathfrak{i}$. First, $0 \in \mathfrak{i}$ because $0 = a - a \in I$. Hence $-a = 0 - a \in \mathfrak{i}$ for $a \in \mathfrak{i}$. We note that all multiples of $a \in \mathfrak{i}$ are also in $\mathfrak{i}$, since it is of the form $a + \cdots + a$ or $(-a) + \cdots + (-a)$. Hence it is an ideal. We distinguish two cases. If $\mathfrak{i} = \{0\}$, then $d = 0$.

Suppose $\mathfrak{i} \neq \{0\}$; then there is a positive integer $a \in \mathfrak{i}$, since $\pm a \in \mathfrak{i}$ for $a \in \mathfrak{i}$. Hence $\varnothing \neq X = \mathfrak{i} \cap \mathbb{N} \subset \mathbb{N}$ and by Theorem 1.2, we may take the least natural number d in $\mathfrak{i}$; then we have $\{kd \mid k \in \mathbb{Z}\} \subset \mathfrak{i}$. Conversely, let m be an integer in $\mathfrak{i}$. Then we apply the Euclidean division (1.2) in the form:

$$m = qd + r \quad q, r \in \mathbb{Z}, \quad 0 \leq r < d. \tag{1.4}$$

Since $m, d \in \mathfrak{i}$, we have $r = m - qd \in \mathfrak{i}$. $r > 0$ contradicts the minimality of d, whence $r = 0$. This yields the reverse inclusion, and (1.3) follows, i.e. $\mathfrak{i}$ is generated by one element. $\square$

Remark 1.1. We remark that Theorem 1.5 amounts to a general theorem to the effect that a Euclidean domain is a principal ideal domain (cf. Exercise 16), specified to the ring of integers. We now have the more proper

Euclidean division (1.4) than (1.2). The integer q is called the quotient and r is called the residue. The residue in the range $[0, q)$ is called the least non-negative residue modulo d. It can be taken in the range $\left[-\frac{d}{2}, \frac{d}{2}\right]$, in which case it is called the absolutely least residue.

Let $a_1, a_2, \cdots, a_m \in \mathbb{Z}$; then

$$i := \{k_1 a_1 + \cdots + k_m a_m \mid k_1, \cdots, k_m \in \mathbb{Z}\}$$

is an ideal, which we often write $(a_1, \cdots, a_m)$. By the theorem above, there is a non-negative integer d, called the **greatest common divisor** of $a_1, \cdots, a_m$ such that

$$i = d\mathbb{Z} \ (= \{kd \mid k \in \mathbb{Z}\})$$

and so

$$(d) = (a_1, \cdots, a_m) \text{ or simply } d = (a_1, \cdots, a_m).$$

We note that $d = 0$ if and only if $a_1 = \cdots = a_m = 0$. Suppose $d \neq 0$; then d is the largest natural number that divides all of $a_1, \cdots, a_m$. For, $a_i \in i = (d)$ means that $d \in i$ divides all a_i. Largest because d being of the form $d = k_1 a_1 + \cdots + k_m a_m$, any divisor of $a_1, \cdots, a_m$ divides d.

If $d = 1$, we say $a_1, \cdots, a_m$ are **relatively prime.**

Let us state all these as a theorem.

Theorem 1.6. *Let $a_1, \cdots, a_m$ be integers and let d be their greatest common divisor. Then there are integers $k_1, \cdots, k_m$ such that*

$$k_1 a_1 + \cdots + k_m a_m = d. \tag{1.5}$$

In particular, if $a_1, \cdots, a_m$ are relatively prime, then $d = 1$ in equation (1.5).

Remark 1.2. In terms of ideals in a general domain, the gcd of two ideals $\mathfrak{a}$ and $\mathfrak{b}$ is the ideal $\mathfrak{a} + \mathfrak{b}$ generated by $\mathfrak{a}$ and $\mathfrak{b}$, i.e. $\mathfrak{a} + \mathfrak{b} = \{a + b \mid a \in \mathfrak{a}, b \in \mathfrak{b}\}$. Similarly for any finitely generated ideals.

Let us mention the **Euclidean algorithm** which gives the greatest common divisor explicitly. Let $a, b \in \mathbb{Z}$ and suppose $ab \neq 0$. Then with $r_1 = |a|, r_2 = |b|$, using (1.4), we find the sequence of natural numbers $r_1, r_2, \cdots$, which in matrix notation reads:

$$\begin{pmatrix} r_k \\ r_{k+1} \end{pmatrix} = \begin{pmatrix} q_k & 1 \\ 1 & 0 \end{pmatrix} \begin{pmatrix} r_{k+1} \\ r_{k+2} \end{pmatrix}, \quad 0 < r_{k+1} < r_k, \ k = 1, 2, \cdots, m, \ r_{m+2} = 0,$$

where q_i, r_i are integers. Hence

$$\binom{r_1}{r_2} = \begin{pmatrix} q_1 & 1 \\ 1 & 0 \end{pmatrix} \cdots \begin{pmatrix} q_m & 1 \\ 1 & 0 \end{pmatrix} \binom{r_{m+1}}{0}$$

$$= M \binom{r_{m+1}}{0}, \text{ say,} \tag{1.6}$$

where the entries of the matrix M are integers and the determinant $\det M$ is $(-1)^m$. Therefore M^{-1} is also an integral matrix, and so

$$\binom{r_{m+1}}{0} = M^{-1}\binom{r_1}{r_2}, \quad M^{-1} = \begin{pmatrix} x_1 & x_2 \\ * & * \end{pmatrix}, \text{ say.} \tag{1.7}$$

By (1.6), d divides r_1 and r_2 if and only if d divides r_{m+1}, and we have

$$(a, b) = r_{m+1} = x_1|a| + x_2|b|.$$

If a non-zero integer a divides an integer b, i.e. $b = an$ for some integer n, we write

$$a \mid b,$$

and we call a a divisor of b, and b a multiple of a and the negation is expressed as $a \nmid b$.

In connection with divisibility, the unit -1 can be dispensed with and we shall mainly restrict to positive divisors. For example, in the case of summation index, $d \mid n$ usually means that d runs over all natural numbers dividing n.

Theorem 1.7. *Let $a\,(\neq 0), b, c$ be integers. Assume $a \mid bc$ and $(a, b) = 1$. Then $a \mid c$.*

Proof. Take integers x, y such that $xa + yb = 1$. Then $c = xac + ybc = a(xc + y(bc)/a)$, and $xc + y(bc)/a$ is an integer. This means c is divisible by a. $\qquad\qquad\square$

Exercise 3. We define the relation $\sim$ in $\mathbb{Z} \times (\mathbb{Z} - \{0\})$ as follows. Two elements $(a_1, b_1) \sim (a_2, b_2)$ if $a_1b_2 - a_2b_1 = 0$. Prove that this is an equivalence relation and describe equivalence classes. We denote the equivalence class containing (a_1, b_1) by a_1/b_1 or $\frac{a_1}{b_1}$. Then, defining addition and multiplication by $a_1/b_1 + a_2/b_2 = (a_1b_2 + a_2b_1)/(b_1b_2)$, $(a_1/b_1)(a_2/b_2) = (a_1a_2)/(b_1b_2)$ prove that $\big(\mathbb{Z} \times (\mathbb{Z} - \{0\})\big)/\sim$ forms a field—the **rational number field**— $\mathbb{Q}$ in the sense of Definition 1.3.

Definition 1.4. As in Exercise 3 we may construct a field K of fractions from an integral domain R. The field of fractions is called a **quotient field** with zero $0/1$ and unity $1/1$ and the elements of R are given in the form $a/1$ with $a \in R$ and is regarded as embedded in K.

1.2   Elements of group theory

In this section we shall collect basic facts in group theory with many exercises of elementary nature, which are far-ranging yet related to group-structure. By solving them, the reader can make a firm grasp of basic facts in various fields of mathematics.

Definition 1.5. Let G be a non-empty set. A **binary operation** is a mapping from $G \times G$ into G which is often denoted by $a \cdot b$ (usually written as ab and called the product of a and b) for two elements $a, b \in G$. If $(G, \cdot)$ satisfies the following conditions, then G is called a **group**,
(i) associative law holds: For $a, b, c \in G$, $(ab)c = a(bc)$, i.e. there is no difference from which two we start and so we may write abc.
(ii) The identity element e exists satisfying $ae = ea = a$ for any $a \in G$. This means that there is a neutral state.
(iii) For each $a \in G$, its inverse exists such that

$$aa^{-1} = a^{-1}a = e.$$

This signifies the Yin-Yan principle to the effect that if an action followed by its reaction results in the neutral state.

If in addition to above conditions,
(iv) the commutative law $ab = ba$ holds,
then G is called an **Abelian group** and the operation is often denoted by $+$.

If G satisfies only condition (i), it is called a **semi-group**.

If a subset H of G forms a group with respect to the operation in G, it is called a **subgroup** of G. The trivial subgroup consisting of the identity element is denoted by 1 (or by 0 if the operation is written additively).

If a group G consists of finite number of elements, G is called a finite group and its cardinality is called the **order** and denoted by $|G|$.

Indeed, this last subgroup terminology is applicable to any algebraic structure.

If a subset of an algebraic system $\mathcal{A}$ with some operations forms the same algebraic system with respect to the same operations in $\mathcal{A}$, it is called a sub-system.

Exercise 4. Prove that a subset $H \neq \varnothing$ of G is a subgroup of G if and only if

(i) $a, b \in G$, then $ab \in G$ and (ii) $a \in G$, then $a^{-1} \in G$.
Prove that these conditions may be replaced by
(iii) If $a, b \in G$, then $ab^{-1} \in G$.

Definition 1.6. In a group G, for a given subset $X \neq \varnothing$ of G, the smallest subgroup H containing X is called the **subgroup generated by** X and denoted by $< X >$, X is then called a **generator** of $H = < X >$. If the set X is finite, then $H = < X >$ is called **finitely generated** (abbreviated by **f.g.**). In particular, a subgroup generated by one element a is called a cyclic subgroup and denoted by $< a >$. If G is generated by one element a, then $G = < a >$ is called a **cyclic group** and a is called a **generator** of G. If $< a >$ is a finite group, we often denote its order by $o(a)$ and refer to it as the **order** of a.

This notion of generators applies to other algebraic systems, too: **In an algebraic system** $\mathcal{A}$, **the smallest sub-system that contains a given set** $X \subset \mathcal{A}$ **is given as the intersection of all sub-systems of** $\mathcal{A}$ **that contain** X (since $\mathcal{A}$ itself is one of them, the intersection is non-empty). It consists of all the elements that arise by applying the operations in $\mathcal{A}$ to the elements in $\mathcal{A}$. For example, if $\mathcal{A}$ is a vector space over $\mathbb{R}$ and $X \subset \mathcal{A}$, then the subspace generated by X is all finite linear combinations of elements in $\mathcal{A}$ with real coefficients.

Proposition 1.2. *(i) Let G be a group and let H and K be subgroups of G. Then $HK = \{hk \mid h \in H, \ k \in K\}$ is a subgroup of G if and only if $HK = KH$. (ii) In the setting of (i), if $H \lhd G$, then HK is a subgroup of G, where $H \lhd G$ means $a^{-1}Ha = H$ for all $a \in G$ (Definition 1.8).*

Proof. (i) Suppose HK is a subgroup of G. Then since $HK \supset K, H$, we have $HK \supset KH$. For any $k \in K$, $h \in H$, we have $k^{-1}h^{-1} \in KH \subset HK$. Hence we may write $k^{-1}h^{-1} = h'k'$, $h' \in H$, $k' \in K$. Hence $hk = \left(k^{-1}h^{-1}\right)^{-1} = (h'k')^{-1} = (k')^{-1}(h')^{-1} \in KH$, i.e. $HK \subset KH$. Therefore $HK = KH$.

Conversely, if $HK = KH$, then for any $h_1 k_1, \ h_2 k_2 \in HK = KH$, we have $k_1 h_2 = h'_2 k'_1 \in HK$. Hence $(h_1 k_1)(h_2 k_2) = h_1 h'_2 k'_1 k_2 \in HK$. Since H is a subgroup of G, we have for any $k \in K$, $(Hk)^{-1} = Hk^{-1} \in HK$ since K is a subgroup. Hence HK is a subgroup of G.

(ii) For any $a, \ b \in K$, using the fact that $a^{-1}Ha = H$ we obtain $Ha(Hb)^{-1} = HaHb^{-1} = Haa^{-1}Hab^{-1} = HHab^{-1} = Hab^{-1} \in HK$. Hence HK is a subgroup of G. $\qquad\square$

Question 1. Prove that $G \lhd G$.

Proof. For any $a \in G$, we have $a^{-1}Ga \subseteq G$. Conversely, any $x \in G$ can be written down as $exe = a^{-1}axa^{-1}a$, which belongs to $a^{-1}Ga$. Hence $a^{-1}Ga = G$.

Second proof. Let $0_G : G \to G$; $0_G(a) = e$ be the zero map, which is a homomorphism. Then $G = \text{Ker } 0_G \lhd G$ (cf. Proposition 1.4).

Question 2. For any $a \in G$, prove that aG is a subgroup of G.

Proof. By Question 1, $G \lhd G$, and so by Proposition 1.2, (i), we have $aG = \{a\} G$ is a subgroup of G.

Proposition 1.3. *Any subgroup of a cyclic group G is again cyclic. If G is finite, say $|G| = q$, then there are $\varphi(q)$ generators of G, where $\varphi(q)$ is the Euler function defined by (1.25).*

Proof. Let $G = <a> \neq 1$. Any subgroup $H \neq 1$ of G consists of elements of the form $a^m, m \in \mathbb{Z}$. Let

$$d = \min\{m \in \mathbb{N} | a^m \in H\}.$$

Then d must divide all m for which $a^m \in H$. For writing

$$m = dm' + r, \ 0 \leq r < d,$$

we have $a^m = (a^d)^{m'} a^r$, so that $a^r \in H$ and $r = 0$. Hence any $a^m \in H$ is of the form $(a^d)^{m'} \in <a^d>$, whence $H = <a^d>$.

Now suppose G is finite and let the order of a (which is the order of G) be q, where the order of an element a is the smallest natural number such that $a^k = 1$. The subgroup $H = <a^d>$ coincides with $G = <a>$ if and only $a \in <a^d>$, i.e. there is an integer m' such that

$$dm' \equiv 1 \, (\text{mod } q),$$

which is true if and only if $(d, q) = 1$. Since there are $\varphi(q)$ such d's, the second assertion follows.   □

Cf. Proposition 1.5 and for another proof cf. Exercise 23.

Example 1.1. Let μ_q denote the set of all q-th roots of 1:

$$\mu_q = \{z \in \mathbb{C} | z^q = 1\}.$$

Then μ_q is a cyclic group generated e.g. by $e^{2\pi i/q}$—piervotnyi koren'.

Note that we may also prove this directly as follows. Clearly, $e^{2\pi i/q}$ is a generator of μ_q and $\mu_q = \{e^{2\pi i a/q} | a = 0, 1, \cdots, q-1\}$. For another member $e^{2\pi i d/q}$ to be a generator, it is necessary and sufficient that $e^{2\pi i/q} \in\; < e^{2\pi i d/q} >$, i.e. that there is an integer x such that $dx \equiv 1 \pmod q$. The last holds if and only $(d, q) = 1$ since then there are integers x, y such that $dx + qy = 1$.

Example 1.2. The group of residue classes modulo q is an additive cyclic group of order q generated by $1 + q\mathbb{Z}$. The group of reduced residue classes modulo a prime is a cyclic group generated by a primitive root (cf. Example 1.8).

1.3 Homomorphisms

We begin by

Definition 1.7. Let G and G' be two groups. f is called a **homomorphism** if it satisfies $f(ab) = f(a)f(b)$.

Proposition 1.4. *Let e and e' be the identity of G and G', respectively. Then*
(i) *the identity maps on the identity*

$$f(e) = e'.$$

(ii) *the inverse maps on the image of the inverse*

$$f(a^{-1}) = f(a)^{-1} \quad for\, a \in G.$$

(iii) *the division law holds*

$$f(ab^{-1}) = f(a)f(b)^{-1} \quad for\, any\, a, b \in G. \tag{1.8}$$

(iv) *the inverse image* $\mathrm{Ker}\, f = \{a \in G | f(a) = e'\}$ *of the identity $e' \in G'$ is a subgroup of G, called the **kernel** of f.*

Proof. (i) Multiplying by $f(e)^{-1} \in G'$, we have

$$e' = f(e)^{-1}f(e) = f(e)^{-1}f(e)f(e) = f(e).$$

(ii) Proof follows on multiplying $f(a)$ by $f(a)^{-1}$.
(iii) follows from multiplicativity and (ii).
(iv) If $a, b \in \mathrm{Ker}\, f$, then $f(ab^{-1}) = f(a)f(b)^{-1} = e'e' = e'$ and $ab^{-1} \in \mathrm{Ker}\, f$. $\qquad\square$

Exercise 5. For $z = x + iy \in \mathbb{C}$, define its **conjugate** $\bar{z}$ by

$$\bar{z} = x - iy \qquad (1.9)$$

prove that $f : \mathbb{C}^\times \to \mathbb{C}^\times$; $f(z) = \bar{z}$ is a homomorphism and derive that

$$\overline{\left(\frac{z_1}{z_2}\right)} = \frac{\overline{z_1}}{\overline{z_2}}$$

for $z_2 \neq 0$.

Proof.   Homomorphicity reads

$$\overline{z_1 z_2} = \overline{z_1}\, \overline{z_2} \qquad (1.10)$$

and as can be readily checked,

$$\overline{\left(\frac{z_1}{z_2}\right)} = f(z_1 z_2^{-1}) = f(z_1) f(z_2^{-1}) = f(z_1) f(z_2)^{-1} = \overline{z_1}\, \overline{z_2}^{-1} = \frac{\overline{z_1}}{\overline{z_2}}. \qquad \square$$

Exercise 6. Choose $G = G' = \mathbb{C}^\times$ and let $f(z) = |z|$. Then check that f is a homomorphism and prove that

$$\left|\frac{z_1}{z_2}\right| = \frac{|z_1|}{|z_2|} \qquad (1.11)$$

if $z_2 \neq 0$.

Solution.   First we prove the homomorphicity of f, or as is more well-known, **multiplicativity**:

$$|z_1 z_2| = |z_1||z_2|. \qquad (1.12)$$

First proof depends on (1.10) and

$$|z|^2 = z\bar{z}. \qquad (1.13)$$

Using these, we have trivially,

$$|z_1 z_2|^2 = z_1 z_2 \overline{z_1 z_2} = z_1 z_2 \overline{z_1}\,\overline{z_2} = z_1 \overline{z_1} z_2 \overline{z_2} = |z_1|^2 |z_2|^2 = (|z_1||z_2|)^2.$$

Second proof uses the **Lagrange identity**

$$|\mathbf{z_1}|^2 |\mathbf{z_2}|^2 = |\mathbf{z_1} \cdot \mathbf{z_2}|^2 + |\mathbf{z_1} \times \mathbf{z_2}|^2, \qquad (1.14)$$

where in general $\mathbf{z} = \begin{pmatrix} x \\ y \end{pmatrix}$ is the vector corresponding to $z = x + iy$.

Lagrange used (1.14) to prove that the product of two integers which are the sums of two squares is again a sum of two squares; e.g. $2 \times 5 = 10$. It also implies the Cauchy-Schwarz inequality (1.21) below.

Now (1.8) reads $|z_1||z_2^{-1}| = |z_1||z_2|^{-1}$, which is (1.11).

Exercise 7. It is appropriate at this point to introduce the complex number field $\mathbb{C}$ as the direct product $\mathbb{R}^2$ with ordinary addition and special multiplication

$$\mathbf{z}_1 * \mathbf{z}_2 = \begin{pmatrix} \bar{\mathbf{z}}_1 \cdot \mathbf{z}_2 \\ \bar{\mathbf{z}}_1 \times \mathbf{z}_2 \end{pmatrix} = \begin{pmatrix} x_1 x_2 - y_1 y_2 \\ x_1 y_2 + x_2 y_1 \end{pmatrix}, \ \mathbf{z}_j = \begin{pmatrix} x_j \\ y_j \end{pmatrix}, \tag{1.15}$$

which corresponds to the multiplication $z_1 z_2$, with $z_j = x_j + i y_j, j = 1, 2$.

Exercise 8. We denote the differentiation (operator) by D, so that $Df = f'$. From the differentiation law of the product

$$D(fg) = Dfg + fDg \tag{1.16}$$

and the condition

$$Dc = 0, \tag{1.17}$$

c being a constant, derive that

$$D\frac{f}{g} = \frac{Dfg - fDg}{g^2} \tag{1.18}$$

provided that $g(x) \neq 0$ in the neighborhood of the point in question.

Solution. First we note that $D\left(\frac{1}{g}\right) = -\frac{Dg}{g^2}$. Then apply this and (1.18) to $D\left(f\frac{1}{g}\right)$, whence (1.18).

We note that although D is not a homomorphism, we may still apply the same argument as in Exercises 5 and 6 to derive (1.18). Any linear operator on the algebra A over a ring R is called a differential operator if it satisfies (1.16). All differential operators over A forms a Lie algebra over R with respect to a suitable product.

Exercise 9. Let G, G' be additive groups with G' equipped with ordering $\leq$.

Suppose f satisfies the inequality

$$f(x + y) \leq f(x) + f(y).$$

Then derive that

$$\max\{f(x) - f(y), f(y) - f(x)\} \leq f(x + y).$$

Exercise 10. Prove the **triangular inequality**

$$|z_1 + z_2| \leq |z_1| + |z_2| \tag{1.19}$$

and derive from this and Exercise 9

$$||z_1| - |z_2|| \leq |z_1 + z_2|. \tag{1.20}$$

Solution. The first proof rests on (1.13) and (1.12).

The second proof follows from the Cauchy-Schwarz inequality:

$$|\mathbf{z}_1 \cdot \mathbf{z}_2| \leq |\mathbf{z}_1||\mathbf{z}_1|. \tag{1.21}$$

(1.20) follows from (1.19) and Exercise 9.

1.4   Quotient groups

Example 1.3. Let H be a subgroup of G. Define a relation $\sim$ between elements x, y in G: $x \sim y$ if there exists an $a \in H$ such that $y = ax$. Then one can prove that this is an equivalence relation. The equivalence class containing $x \in G$ is of the form $Hx = \{ax|a \in G\}$ and is called the **right coset**, whose cardinality is $|H|$. If $H \backslash G = \{Hx_\nu | \nu \in N\}$, i.e. $G = \bigcup_\nu Hx_\nu$ (disjoint), is the right coset decomposition of G, then $\{x_\nu\}$ is called a **complete set of right representatives** of G with respect to H. In a similar way, we may consider the **left coset** decomposition of G with respect to H: $G/H = \{x_\mu H | \mu \in M\}$, with x_μ a complete set of left representatives of G with respect to H.

Theorem 1.8. *The following two conditions are equivalent.*
$\{x_\nu\}$ *forms a complete set of right representatives:* $H \backslash G = \{Hx_\nu\}$.
$\{x_\nu^{-1}\}$ *forms a complete set of left representatives:* $G/H = \{x_\nu^{-1}H\}$.

Proof.   Suppose the first is true. It is enough to show that $\{x_\nu^{-1}H\} = G/H$, i.e. that $x_\nu^{-1}H \neq x_\mu^{-1}H$ if $x_\nu = x_\mu$ and that any $a \in G$ is contained in some $x_\nu^{-1}H$. The first because if $x_\nu^{-1}H = x_\mu^{-1}H$, then

$$Hx_\nu = \left(x_\nu^{-1}H\right)^{-1} = \left(x_\mu^{-1}H\right)^{-1} = Hx_\mu,$$

whence $x_\nu = x_\mu$.

The second is true because for any $a \in G$, there exists an x_ν such that $a^{-1} \in Hx_\nu$, which means that there exists a $y \in H$ such that $a^{-1} = yx_\nu$. Hence $a = x_\nu^{-1}y^{-1} \in x_\nu^{-1}H$.   □

Theorem 1.8 asserts that the cardinality of right cosets and left cosets are equal. If it is finite, we denote it by

$$(G : H)$$

and call it the **index** of H in G. Since each coset has $|H|$ elements, we have an important identity

$$|G| = (G : H)|H|. \tag{1.22}$$

This is also valid for $|G| = |H| = \infty$ by interpreting both sides to be ∞. An example is

$$|\mathbb{Z}| = (\mathbb{Z} : q\mathbb{Z})|q\mathbb{Z}|,$$

where in this case $(\mathbb{Z} : q\mathbb{Z})$ is the order of the additive group $\mathbb{Z}/q\mathbb{Z}$ of residue classes mod q, $q \in \mathbb{N}$ (see below).

Since each group contains the trivial group consisting of the identity element only, we have $|G| = (G : \{1\})$.

For finite groups, the following corollary is fundamental.

Corollary 1.1. (Lagrange) *Let G be a finite group and let H be its subgroup. Then the order of H divides that of G:*

$$|H| \,\big|\, |G|.$$

Corollary 1.1 is used, e.g., in the following context.

Recall Definition 1.6 of the cyclic subgroup $< a >$ generated by a in a group G. It is the set of all powers of a:

$$\langle a \rangle = \{a^m | m \in \mathbb{Z}\}.$$

There are two cases $|\langle a \rangle| = \infty$ or $|\langle a \rangle| < \infty$. In the latter case, we have the **generalized Fermat's little theorem**:

$$a^{|G|} = 1. \tag{1.23}$$

This follows from the fact that $|< a >| = \min\{m \in \mathbb{N} | a^m = 1\}$. Indeed, since there are only finitely many different powers of a, we must have $a^k = a^l$, whence we may conclude that $X = \{m \in \mathbb{N} | a^m = 1\} \neq \varnothing$. Choosing $l = \min X$, we may apply the same argument as in the proof of Theorem 1.5 to conclude that l divides all members of X and that all elements $1, a, \ldots, a^{l-1}$ are distinct, so that $|< a >| = l$. By Lagrange's theorem, $a^{|G|} = \left(a^{|<a>|}\right)^{\frac{|G|}{|<a>|}} = 1$.

Definition 1.8. A subgroup N of a group G is called a **normal** subgroup, denoted $N \lhd G$ if the right and left cosets coincide: $aN = Na$ or closed under the inner automorphism $a^{-1}Na = N$. Let G be a group and let N be its normal subgroup. In the quotient G/N consisting of cosets, we introduce multiplication:

$$aNbN = abN, \quad a, b \in G. \tag{1.24}$$

We need to verify that (1.24) is well-defined because multiplication is to be a map from the Cartesian product $G/N \times G/N$ into G/N and a coset has many representatives. This is easy to check. If $a'N = aN$ and $b'N = bN$, then since $a' = an_1, b' = bn_2$, it follows that $a'b' = an_1bn_2$ and using the normality of N, $an_1bn_2 = abn_3n_2$ for some $n_3 \in N$, so that $a'Nb'N = aNbN$. It is easy to check that with this well-defined multiplication, the quotient G/N forms a group, called the **quotient group**.

Definition 1.9. Suppose R is a ring and $\mathfrak{i}$ is its ideal (recall Definition 1.3). Then we form the quotient group as an additive group in the sense of Definition 1.8. We may introduce the well-defined multiplication between two elements $a + \mathfrak{i}$ and $b + \mathfrak{i}$ of $R/\mathfrak{i}$ by $(a + \mathfrak{i})(b + \mathfrak{i}) = ab + \mathfrak{i}$. It is easy to check that $R/\mathfrak{i}$ forms a ring, called the **quotient ring**.

Example 1.4. Let $q \in \mathbb{N}$. Then we form the quotient ring $\mathbb{Z}$ modulo the ideal $(q) = q\mathbb{Z}$: $\mathbb{Z}/q\mathbb{Z} = \{a + q\mathbb{Z} | a \in \mathbb{Z}\}$. The unit group $(\mathbb{Z}/q\mathbb{Z})^{\times}$ consists of those cosets whose elements are relatively prime to q, called **reduced residue classes** mod q. Indeed, by Theorem 1.6, if $(a, q) = 1$, then there are integers x, y such that $ax + qy = 1$, whence $x + q\mathbb{Z} = (a + q\mathbb{Z})^{-1}$. The unit group is usually called the **group of reduced residue classes** mod q:

$$(\mathbb{Z}/q\mathbb{Z})^{\times} = \{a + q\mathbb{Z} | (a, q) = 1\}.$$

The order of this group, denoted by $\varphi(q)$, is called **Euler's function**:

$$\varphi(q) = |(\mathbb{Z}/q\mathbb{Z})^{\times}|, \tag{1.25}$$

i.e.

$$\varphi(q) = \sum_{\substack{k \leq q \\ (k,q)=1}} 1 = \text{the number of integers } k,\ 1 \leq k \leq q,\ \text{coprime to } q.$$

$$\tag{1.26}$$

Proposition 1.5. *Another interpretation of the Euler function $\varphi(k)$ is that it is the number of generators of a cyclic group of order k.*

This is the same as Proposition 1.3.

Proposition 1.6. *Let $G = <a>$ be a cyclic group of order n. Then for every divisor d of n there exists a unique subgroup of order d with $\varphi(d)$ generators and we have the identity*

$$\sum_{d|n} \varphi(d) = n. \tag{1.27}$$

Proof. Recall from Proposition 1.3 that there are $\varphi(n)$ generators in G including a.

Now for each divisor d of n, the element $a^{n/d}$ generates a subgroup H_d of order d.

There are $\varphi(d)$ generators of H_d and H_d $(d|n)$ *exhaust* all the elements of G: $G = \cup_{d|n} H_d$ (disjoint). Hence (1.27) follows as the cardinality of both sides. □

This is an example of the **exhaustion principle**. For another example, see Example 1.1.

Theorem 1.9. *Let G be a group. A subgroup N of G is normal if and only if N is a kernel of a homomorphism from G into another group G'.*

Proof. Suppose $N \lhd G$. Then we may form the quotient group G/N whose multiplication being given by $(aN)(bN) = abN$. This is well-defined because if $a'N = aN$, $Nb' = Nb$, then $a^{-1}a' \in N$, and so $b^{-1}a^{-1}a'b' \in b^{-1}Nb' = b^{-1}Nb = N$. Hence $a'b'N = abN$.

Let $\pi_N : G \to G/N$, $\pi_N(a) = aN$ be the canonical projection. Then $\pi_N(ab) = abN = aNbN = \pi_N(a)\pi_N(b)$ and π_N is a homomorphism $G \to G/N$ and $\operatorname{Ker} \pi_N = N$.

Now, suppose a homomorphism $f : G \to G'$ is given. Let $N = \operatorname{Ker} f$. Then for any $a \in G$, $b \in aN \iff a^{-1}b \in N \iff f(a^{-1}b) = 1_{G'} \iff f(a)^{-1}f(b) = f(a^{-1}b) = 1_{G'} \iff f(a) = f(b) \iff f(ba^{-1}) = f(b)f(a)^{-1} = 1_{G'} \iff ba^{-1} \in N \iff b \in Na$. Hence $aN = Na$, i.e. $N \lhd G$. $\qquad \square$

Theorem 1.10. (Homomorphism Theorem) *If f is a homomorphism from a group G into a group G'. Then there exists an isomorphism $\bar{f} : G/N \to \operatorname{Im} f$ such that the diagram commutes, where N is $\operatorname{Ker} f \lhd G$.*

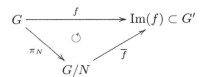

The first homomorphism theorem.

Proof. The mapping

$$\bar{f}(aN) = f(a), \qquad a \in G$$

is a well-defined homomorphism. For if $aN = bN$, then $b^{-1}aN = N$ and $1_{G'} = \bar{f}(N) = \bar{f}(b^{-1}aN) = f(b^{-1}a) = f(b)^{-1}f(a)$, whence $f(a) = f(b)$ and $\bar{f}$ is well-defined.

Secondly, since $aNbN = abN$, it follows that $\bar{f}(aNbN) = \bar{f}(abN) = f(ab) = f(a)f(b) = \bar{f}(aN)\bar{f}(bN)$, and $\bar{f}$ is a homomorphism. $\bar{f}$ is onto on $\operatorname{Im} \bar{f} \subset G'$.

Finally, since $\text{Ker } \bar{f} = N$, it follows that $f(a) = f(b)$ implies $\bar{f}(aN) = \bar{f}(bN)$ and

$$\bar{f}((bN)^{-1}(aN)) = \bar{f}(bN)^{-1}\bar{f}(aN) = f(b)^{-1}f(a) = 1_{G'},$$

so that $(bN)^{-1}aN \in N$, and $aN = bN$. Hence $\bar{f}$ is one-to-one, so that $\bar{f}$ is an isomorphism $G/N \to \text{Im }\bar{f}$. Since $\bar{f} \circ \pi_N = f$, the diagram is commutative.   □

Corollary 1.2. *An infinite cyclic group is isomorphic to $\mathbb{Z}$ and a finite cyclic group is isomorphic to $\mathbb{Z}/m\mathbb{Z}$ $(1 < m \in N)$.*

Proof.   Let $G = <a> = \{na | n \in \mathbb{Z}\}$ and define $f : \mathbb{Z} \to G$ by $f(n) = na$ $(n \in \mathbb{Z})$. Then f is a surjective homomorphism. Since $\text{Ker } f$ is an (additive) subgroup of $\mathbb{Z}$, it must be of the form $m\mathbb{Z}$ with $m \in N \cup \{0\}$ in view of Theorem 1.5.

The homomorphism theorem implies that

$$G \cong \mathbb{Z}/m\mathbb{Z}.$$

The case $m = 0$ or 1 is the case of an infinite cyclic group.   □

Theorem 1.11. (The second homomorphism theorem) *Let H and N be subgroups of G and let $N \lhd G$. Then $H \cap N \lhd H$ and we have the isomorphism*

$$H/H \cap N \cong HN/N$$

under the correspondence $a(H \cap N) \longleftrightarrow aN$.

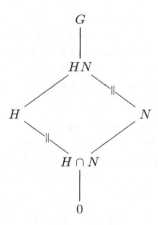

The second homomorphism theorem.

Proof. The map $f(a) = aN$, $f : H \to HN/N$ is a surjective homomorphism. Since $\operatorname{Ker} f = \{a \in H | aN = N\} = H \cap N$, the assertion follows from Theorem 1.10. $\qquad\square$

Exercise 11. Prove that the groups of order 4 are either cyclic or isomorphic to the **Klein four group** (die Kleinsche Viertergruppe) V_4 which is the point group of a rectangle with unequal sides and can be identified with the subgroup
$$\{e, \, (1,2)(3,4), \, (1,3)(2,4), \, (1,4)(2,3)\}$$
of the symmetric group S_4 of 4 symbols. Cf. Proposition 2.3 as its role as a Galois group.

Solution. From the passage after Corollary 1.1 it follows that the orders of all the elements of the group G of order 4 are divisors of 4, i.e. $1, 2, 4$. The order 1 element is the identity, while if there is an element of order 4, it is a generator of G and G is a cyclic group, which is isomorphic to $\mathbb{Z}/4\mathbb{Z}$ by Corollary 1.2. If there is no element of order 4, then there must be at least two elements a, b of order 2. Then $ab \in G$, which is different from a, b, whence $G = \{1, a, b, ab\}$. Now since $ba \in G$, it must coincide with ab. Hence G is an Abelian group. Indeed, this fact is a special case of the fact that any group of order p^2, p being a prime is cyclic. Hence in this case, G is the direct product of $< a >$ and $< b >$, a $(2,2)$-elementary Abelian group. It turns out that this G coincides with V_4.

1.5   Rings and fields

According to Definition 1.3, a non-empty set R is called a **ring** if on it there are defined two operations, addition and multiplication under which R is an additive Abelian group and multiplicative semi-group and the distributive hold true. We give a concrete description when R is a field, i.e. $R^{\times} = R - \{0\}$ forms a multiplicative group.

Definition 1.10. Let K be a set in which for any two elements $a, b \in K$ their sum $a + b$ and product $a \cdot b$, which we denote by ab, are uniquely defined. If the following rules of operation are satisfied, K is called a **commutative field:** In the following, a, b, c stand for arbitrary elements of K.

(i) (associative law) $(a + b) + c = a + (b + c)$, $(ab)c = a(bc)$.
(ii) (existence of the identity) There is an element $0 \in K$ such that $a + 0 = 0 + a = a$ and an element $1 \in K$ such that $1a = a1 = a$.

(iii) (existence of the inverse element) There is an element $a' \in K$ such that $a + a' = a' + a = 0$. We denote a' by $-a$ and write $b - a$ for $b + (-a)$ (subtraction). For each non-zero element a of K, there is a^{-1} such that $aa^{-1} = a^{-1}a = 1$.

(iv) (distributive law) $a(b + c) = ab + ac$, $(a + b)c = ac + bc$.

(v) (commutative law) $a + b = b + a$, $ab = ba$.

Exercise 12. State the reason why associative law and distributive are to be added while commutative law is rather additional.

Solution. Since both addition and multiplication are binary operations, we have to distinguish which two of the three we combine first and if the two combinations do not coincide, the operation would be too cumbersome and will yield little. Distributive law is more optional and sometimes this is replaced by other formulas (e.g. in the case of algebras). The law itself naturally arises because there are two operations and when we conduct two operations on three elements, there are two different cases. Since anyway, rings and fields are abstract number systems, we model on the latter and assume distributive law as we have in the number system. Commutative law for addition is naturally added as in the number system while that for multiplication is rarely found in nature. This comes from the fact that natural processes are mostly non-commutative—non-Abelian. Therefore we distinguish commutative and non-commutative algebraic systems with respect to multiplication. More complicated algebraic systems are largely non-Abelian.

Definition 1.11. Let R be a ring. An ideal $\mathfrak{a} \subset R$ is called **maximal** if there is no ideal containing $\mathfrak{a}$, and properly contained in R, i.e. $\mathfrak{a}$ is an ideal such that if

$$\mathfrak{a} \subset \mathfrak{a}' \subset R,$$

then either $\mathfrak{a}' = R$ or $\mathfrak{a}' = \mathfrak{a}$.

Proposition 1.7. *Let E be a non-empty set $\subsetneq R$ and suppose there is an ideal $\mathfrak{a}_0$ of R such that $\mathfrak{a}_0 \cap E = \phi$. Then there exists a maximal ideal $\mathfrak{a}$ such that $\mathfrak{a} \supset \mathfrak{a}_0$ and $\mathfrak{a} \cap E = \phi$.*

For the proof we use **Zorn's lemma**: An **inductive set** has a maximal element, where an inductive set is a partially ordered set each of whose totally ordered subset has an upper bound.

Proof of Proposition 1.7. Let X denote the set of all ideals $\mathfrak{a}$ of R satisfying the conditions

$$\mathfrak{a} \supset \mathfrak{a}_0 \quad \text{and} \quad \mathfrak{a} \cap E = \phi.$$

For any totally ordered subset $\{\mathfrak{a}_\nu\}$ of X,

$$\mathfrak{a} := \bigcup \mathfrak{a}_\nu$$

is an ideal of R satisfying the conditions $\mathfrak{a} \supset \mathfrak{a}_0$ and $\mathfrak{a} \cap E = \phi$, and so $\mathfrak{a}$ is an upper bound of $\{\mathfrak{a}_0\}$. Hence X is an inductive set and so it has a maximal element.

Corollary 1.3. *Suppose R contains the identity 1. Then for any ideal $\mathfrak{a}_0 \subsetneq R$, there exists a maximal ideal $\mathfrak{a} \subset R$ containing $\mathfrak{a}_0$.*

Proof. Take $E = \{1\}$ in the above Proposition. Then it asserts the existence of a maximal ideal $\mathfrak{a} \subsetneq R$ such that

$$\mathfrak{a} \supset \mathfrak{a}_0 \quad \text{and} \quad \mathfrak{a} \cap \{1\} = \phi,$$

the latter condition meaning just that $\mathfrak{a} \subsetneq R$, one of the requirements for $\mathfrak{a}$ to be maximal. $\qquad\square$

A commutative ring R with 1 is called a **simple ring** if the only non-trivial ideal is R itself, i.e. if $\mathfrak{i} \subset R$ is an ideal, then $\mathfrak{i} = 0$ or R.

Exercise 13. Prove that a commutative ring R with 1 is a simple ring if and only if it is a field.

Solution. If R is a field, then for any ideal $\mathfrak{i} \neq 0$, choose $\mathfrak{i} \ni a \neq 0$. Then there is $a^{-1} \in R$ and $1 = a^{-1}a \in \mathfrak{i}$, whence $\mathfrak{i} = R$.

Conversely, if R has no non-trivial ideal save for R itself, then for $R \ni a \neq 0$, the ideal generated by a is non-trivial and must coincide with R. Hence $1 \in (a)$, and so there is a $b \in R$ such that $ab = 1 \in R$. Hence $a^{-1} = b$ exists in R, i.e. R is a field.

Proposition 1.8. *Let R be a commutative ring with 1 and let $\mathfrak{m}$ be an ideal of R. Then $\mathfrak{m}$ is maximal if and only if $R/\mathfrak{m}$ is simple. In particular, if $\mathfrak{m}$ is maximal, then $R/\mathfrak{m}$ is a field.*

Proof. First note that any ideal in $R/\mathfrak{m}$ is of the form $\mathfrak{i}/\mathfrak{m}$, where $\mathfrak{i}$ is an ideal of R containing $\mathfrak{m}$. We exclude the trivial case of $\mathfrak{i} = 0$. If $\mathfrak{m}$ is maximal, then $\mathfrak{i}$ must be R, i.e. $R/\mathfrak{m}$ is simple. Conversely, if $R/\mathfrak{m}$ is simple, again $\mathfrak{i}/\mathfrak{m}$ must be either 0 or R, i.e. either $\mathfrak{m}$ or R, whence $\mathfrak{m}$ is maximal. $\qquad\square$

Definition 1.12. A non-zero element a of a ring R is called a **zero divisor** if there is a non-zero element $b \in R$ such that $ab = 0$. A ring without zero-divisors is called an **integral domain** or simply a domain. An ideal $\mathfrak{p}$ of a commutative ring $\neq \{0\}$ is called a **prime ideal** if $ab \in \mathfrak{p}$ implies $a \in \mathfrak{p}$ or $b \in \mathfrak{p}$. An element that generates a prime ideal is called a **prime**.

Exercise 14. Let R be a ring and let $\mathfrak{a}$ be an ideal of R. Prove that $\mathfrak{a}$ is a maximal ideal if and only if the quotient ring $R/\mathfrak{a}$ is a field and that $\mathfrak{p}$ is a prime ideal if and only if $R/\mathfrak{p}$ is an integral domain.

Solution. $R/\mathfrak{p}$ being an integral domain means that $(a + \mathfrak{p})(b + \mathfrak{p}) = 0$ if and only if $a \in \mathfrak{p}$ or $b \in \mathfrak{p}$, i.e. that $\mathfrak{p}$ is a prime ideal.

Remark 1.3. In a PID, a prime element and an irreducible element are the same by Proposition 1.12 and we may speak of a **prime decomposition**. In this book we are mainly concerned with integral domains (except for the case of finite rings).

Many notions and results in group theory can be naturally extended to those in ring theory.

For two ideals $\mathfrak{a}, \mathfrak{b}$ of R, the **sum** of $\mathfrak{a}$ and $\mathfrak{b}$ is the ideal generated by $\mathfrak{a}$ and $\mathfrak{b}$, i.e. the set of all elements $a + b$, $a \in \mathfrak{a}$, $b \in \mathfrak{b}$:

$$\mathfrak{a} + \mathfrak{b} = \{a + b | a \in \mathfrak{a}, b \in \mathfrak{b}\}$$

and the **product** $\mathfrak{a}\mathfrak{b}$ is defined to be the ideal generated by all the products of elements in $\mathfrak{a}$ and $\mathfrak{b}$:

$$\mathfrak{a}\mathfrak{b} = \{\sum a_i b_i | a_i \in \mathfrak{a}, b_i \in \mathfrak{b}\}.$$

Exercise 15. Prove that $\mathfrak{a}\mathfrak{b}$ is an ideal of R and that

$$\mathfrak{a}\mathfrak{b} \subset \mathfrak{a} \cap \mathfrak{b}. \tag{1.28}$$

An additive group homomorphism $f : R \to R'$ is called a **ring homomorphism** if $f(ab) = f(a)f(b)$.

Theorem 1.12. (Ring homomorphism theorem) *If $f :\to R'$ is a ring homomorphism, then $\mathfrak{a} = \operatorname{Ker} f = \{a \in R | f(a) = 0\}$ forms an ideal of R and $R/\mathfrak{a} \cong \operatorname{Im} f$.*

Theorem 1.13. (Second ring homomorphism theorem) *If S is a subring of R and $\mathfrak{a}$ an ideal of R, then $S \cap \mathfrak{a}$ is an ideal of R and*

$$S/S \cap a \cong (S + \mathfrak{a})/\mathfrak{a}.$$

Proposition 1.9. *Let R be a commutative ring with 1 and let $\mathfrak{a}, \mathfrak{b}$ be ideals of R such that $\mathfrak{a} + \mathfrak{b} = R$. Then*

$$\mathfrak{a}\mathfrak{b} = \mathfrak{a} \cap \mathfrak{b}.$$

Proof. Enough to prove the inverse inclusion of (1.28). There are $a' \in \mathfrak{a}$, $b' \in \mathfrak{b}$ such that $a + b = 1$. Hence for any $x \in \mathfrak{a} \cap \mathfrak{b}$, $x = x1 = xa' + xb' \in \mathfrak{b}\mathfrak{a} + \mathfrak{a}\mathfrak{b} = \mathfrak{a}\mathfrak{b}$. □

Proposition 1.10. *Let R be a commutative ring with 1 and $\mathfrak{a}, \mathfrak{b}$ be ideals of R such that $\mathfrak{a} + \mathfrak{b} = R$. Then*

$$\mathfrak{b}/\mathfrak{a}\mathfrak{b} \cong R/\mathfrak{a}. \tag{1.29}$$

Proof. By Theorem 1.13,

$$\mathfrak{b}/\mathfrak{a} \cap \mathfrak{b} \cong (\mathfrak{a} + \mathfrak{b})/\mathfrak{a},$$

which amounts to (1.29) in view of Proposition 1.9. □

Remark 1.4. Proposition 1.10 means the following multiplicativity of the norm:

Defining $N\mathfrak{a} = (R : \mathfrak{a})$ when it is finite, we have

$$N\mathfrak{a}\mathfrak{b} = (R : \mathfrak{a}\mathfrak{b}) = (R : \mathfrak{b})(\mathfrak{b} : \mathfrak{a}\mathfrak{b}) = N\mathfrak{b}N\mathfrak{a},$$

in the case $\mathfrak{a} + \mathfrak{b} = R$.

Corollary 1.4. *Let R be an integral domain with unity $1 = 1_R$. Let $R_0 = \{n \cdot 1 | n \in \mathbb{Z}\}$ be the subdomain generated by 1, called the **prime domain** of R. Then R_0 is isomorphic to $\mathbb{Z}/p\mathbb{Z}$, p being a prime or 0.*

Proof. The group homomorphism defined in the proof of Corollary 1.2

$$f(n) = n1 \quad (n \in \mathbb{Z})$$

is a ring homomorphism. Hence the ring homomorphism theorem implies

$$R_0 \cong \mathbb{Z}/m\mathbb{Z}.$$

Since R_0 is an integral domain, m is either 0 or a prime p, in the latter of which case, the prime ring $\mathbb{Z}/p\mathbb{Z}$ is the **prime field**

$$\mathbb{F}_p = \mathbb{Z}/p\mathbb{Z} = \{\bar{0}, \bar{1}, \cdots, \overline{p-1}\}. \qquad □$$

Definition 1.13. In the case $m = p$, p a prime, of Corollary 1.4, $R \supset \mathbb{Z}/p\mathbb{Z}$ and R is called **of characteristic** p, written char $R = p$, while if $m = 0$, then $R \supset \mathbb{Z}$ and R is called of characteristic 0, written char $R = 0$. In particular, we may speak of a field K of characteristic p, p being a prime p or 0: $K \supset \mathbb{F}_p$ if char $K = p$ while $k \supset \mathbb{Q} \supset \mathbb{Z}$ if char $K = 0$.

Proposition 1.11. *(i) If* char $R = 0$, *then for* $R \ni a \neq 0$, $\mathbb{Z} \ni n \neq 0$, *we have* $na \neq 0$.

(ii) If char $R = p \neq 0$, *then for* $a \in R$, $pa = 0$. *Also, the map* $f : a \to a^p$ *is an injective automorphism of* R.

Proof.   (i) Since $n \cdot 1 \neq 0$, it follows that for any $a \neq 0$, $na = (n \cdot 1) \cdot a \neq 0$.
(ii) First,

$$pa = (p \cdot 1)a = 0 \cdot a = 0.$$

Secondly, noting that the binomial coefficient $\binom{p}{n} \equiv 0 \,(\mathrm{mod}\, p)$ for $0 < n < p$, we have

$$(a + b)^p = a^p + b^p.$$

Since $(ab)^p = a^p b^p$, it follows that f is a ring homomorphism into itself.
    Finally, Ker $f = \{a | a^p = 0\} = 0$.
    Hence f is injective.   □

Definition 1.14. A domain R is called a **unique factorization domain** (UFD) if

(U.1) every non-zero, non-unit element can be expressed as a product of irreducible elements: $a = p_1 \cdots \cdot p_r$ and up to the order, the expression is unique;

(U.2) If there are two expressions $a = p_1 \cdots \cdot p_r = q_1 \cdots \cdot q_s$, then $r = s$ and changing the order suitably, we have $p_i \approx q_i$.

Proposition 1.12. *For a principal ideal domain (PID)* R, *the following assertions on an element* $p \in R$ *are equivalent.*

a) *p is* **irreducible**, *which means that if* $p = qr$, *then either* $q \in R^\times$ *or* $r \in R^\times$.
b) *If* $p | ab$, *then we have* $p | a$ *or* $p | b$.
c) *$(p) = Rp$ is a prime ideal.*
d) *$(p) = Rp$ is a maximal ideal.*

Proof.   a) $\Rightarrow$ d). Since R is a PID, every ideal of R is principal. Hence, if $R \supset (q) \supset (p)$, then $p = qr$, $r \in R$. Then (a) implies that $q \in R^\times$ or $r \in R^\times$.
If $q \in R^\times$, then $1 \in (q)$ and $(q) = R$.
If $r \in R^\times$, then $(q) \subset (p)$, whence $(q) = (p)$.

$b) \Rightarrow a)$. if $p = ab$, then by $b)$, $p \mid a$ or $p \mid b$. If $p \mid a$, we have $a = pa', a' \in R$, so that $p = pa'b$, $a'b = 1$. Therefore $b \in R^\times$ and $p \approx a$.

Similarly, if $p \mid b$, we have $a \in R^\times$, and $p \approx b$. Hence p is irreducible. $d) \Rightarrow c)$. Assume $ab \in (p)$, and $a \notin (p)$, Then with (a, p) the ideal generated by a and P, we have $(p) \subset (a, p) = Ra + Rp \subset R$. Since (p) is maximal ideal, $(a, p) = R$. there are two elements $x, y \in R$, and $ax + py = 1$.we have $b = b \cdot 1 = abx + bpy \in p$. Similarly, we can prove that if $b \notin (p)$, then $a \in (p)$. □

Exercise 16. Prove that a principal ideal domain is a unique factorization domain.

Solution. Suppose $0 \neq a \notin R^\times$ is not decomposable into the product of irreducible elements. Since $(a) \subsetneq R$, by Corollary 1.3, there exists a maximal ideal $(p_1) \supseteq (a)$, where by Proposition 1.12, p_1 is a prime. Hence $a = p_1 a_1$, $(a) \subset (a_1)$, and by assumption, a_1 is non-decomposable. In the same way, there exists a prime p_2 such that $a_1 = p_2 a_2$. Repeating this process, we get an infinite sequence of ideals $(a) \subset (a_1) \subset (a_2) \subset \cdots$. Let $a = \bigcup \{(a_i) \mid i \in N\} \subset R$. This is an ideal. Since P is a PID, there exists an $x \in R$ such that $a = (x)$, so that $x \in a$ and so there must exist $k \in N$ such that $x \in (a_k)$. Therefore $a = (x) \subseteq (a_k)$, so that $a = (a_k)$, and we get $(a_{k+1}) \subseteq a = (a_k) \subset (a_{k+1})$, a contradiction, whence (U.1) follows. Secondly, if $a = p_1 p_2 \cdots p_r = q_1 q_2 \cdots q_s$, then we get $p_1 \mid q_1 q_2 \cdots q_s$ i.e. p_1 must divide one of $q_1, q_2, \cdots, q_s$. Suppose $p_1 \mid q_1$. Then we must have $p_1 \approx q_1$ and $p_2 \cdots p_r \approx q_2 \cdots q_s$ Repeating this process, we conclude that $r = s$ and $p_i \approx q_i$, completing the proof.

Let R be an integral domain with a function $g : R \to \mathbb{Z}_{\geq 0}$ satisfying $g(a) = 0$ if only if $a = 0$ and for any $a, b \neq 0$, we have $a = bq + r$, $r = 0$ or $g(r) < g(b)$. Such a domain is called a **Euclidean domain**. We can easily show that *a Euclidean domain is a principal ideal domain* (cf. the passage preceding Theorem 1.5). Indeed, for any non-zero ideal $a \subset R$, $0 \neq \exists x_0 \in a$ such that $g(x_0) \in \mathbb{N}$, and so $X = \{g(a) \mid a \in a\} \cap \mathbb{N} \neq \varnothing$. Hence there is a minimal element $g(b)$ in X with $b \in a$. For $\forall a \in a$, $a = bq + r$, $r = a - bq \in a$. Since $g(r) < g(b)$, we must have $g(r) = 0$ so that $r = 0$, and $a = bq$, i.e. $a = (b)$. In particular, the polynomial ring $K[X]$ over a field K is a Euclidean domain with $g(f) = \deg f$, and so it is a PID. By Exercise 16, it is a UFD. Note that any non-zero ideal a is generated by a polynomial $f \in a$ with the minimum degree.

Proposition 1.13. *The polynomial ring $K[X]$ over a field K is a PID, and hence a UFD.*

Definition 1.15. If K is a subfield of a field L, then L is said to be an **extension** field of K and this relation is denoted L/K (read: el over kei). When L is viewed as a vector space over K, the dimension $\dim_K L$ is called the **extension degree** and denoted $[L : K]$. When $[L : K] < \infty$, the extension is called **finite** and if $[L : K] = n$, the extension L/K is called an extension of degree n. Let L/K be an extension and S is a subset of L, then the smallest subfield of L containing $K \cup S$ is called the field **generated** by S, or the field **adjoined** by S over K and denoted $K(S)$. If S is a finite set, then $K(S)$ is called **finitely generated** (in abbreviation: f.g.) over K and if S is a singleton $\{\theta\}$, then $K(\theta) = K(\{\theta\})$ is called a **simple extension**.

Proposition 1.14. *Any extension is obtained as a union of finitely generated extensions, and f.g. extensions are obtained as repetition of simple extensions.*

Proof. Let $K(S)/K$ be an extension. The first assertion means that $K(S) = \cup_F K(F)$, where F ranges over all finite subsets of S. Plainly, $K(S) \supset \cup_F K(F)$. In Exercise 17, it is proved that $\cup_F K(F)$ is a field. Since it contains both K and S, it contains $K(S)$, whence they coincide. To prove the second assertion, we prove a general formula $K(S_1)(S_2) = K(S_1 \cup S_2)$. Since $K(S_1 \cup S_2)$ is a field containing K, S_1, S_2, it must contain $K(S_1)(S_2)$. Conversely, since $K(S_1 \cup S_2) \supset K(S_1)$, it follows that $K(S_1 \cup S_2)$ is a field containing $K(S_1)$ and S_2, so that it contains $K(S_1)(S_2)$. Hence they coincide. $\qquad\square$

Exercise 17.

(i) Prove that $\cup_F K(F)$ is a field, where F ranges over all finite subsets of S.

(ii) If L/K is an extension and $\theta_1, \cdots, \theta_n$ are in L, then the field $K(\theta_1, \cdots, \theta_n)$ consists of the elements of L of the form

$$\frac{f(\theta_1, \cdots, \theta_n)}{g(\theta_1, \cdots, \theta_n)}, \quad f, g \in K[X_1, \cdots, X_n], \quad g(\theta_1, \cdots, \theta_n) \neq 0.$$

Solution. If α, β are in $\cup_F K(F)$, there are finite sets F_1, F_2 such that $\alpha \in F_1, \beta \in F_2$. Hence $\alpha, \beta \in K(F_1 \cup F_2)$. Since $K(F_1 \cup F_2)$ is the field adjoined by $F_1 \cup F_2$, it is contained in $\cup_F K(F)$. Moreover, the sums, differences, products and quotients of α, β are in $F_1 \cup F_2$ and hence in

$\cup_F K(F)$. Hence $\cup_F K(F)$ forms a field. The second assertion can be easily checked.

Theorem 1.14. *A simple extension $K(\theta)$ of a field K is isomorphic to a residue class filed $K[X]/(f)$ of the polynomial ring $K[X]$ with an irreducible polynomial f or to the rational function field $K(X)$. In the former case, θ is a root of the polynomial f.*

Proof.   Let $L = K(\theta)$ and consider the homomorphism

$$\varphi : K[X] \to L;\ \varphi(f) = f(\theta) \quad (f \in K[X]).$$

$\mathrm{Ker}\,\varphi$ is an ideal of $K[X]$ and it is a principal ideal (f), say, by Proposition 1.13. By the ring homomorphism theorem,

$$L \supset \varphi(K[X]) \cong K[X]/(f).$$

Since L is a field, its subring $K[X]/(f)$ must be an integral domain. Hence $f = 0$ or a prime, and so by Proposition 1.12, $f = 0$ or an irreducible polynomial. If f is irreducible, then $\varphi(K[X])$ is a field containing $K(= \varphi(K))$ and $\theta(\varphi(X))$, and $K(\theta) \supset \varphi(K[X])$, whence $K(\theta) = \varphi(K[X]) \cong K[X]/(f)$.

If $f = 0$, then $\varphi(K[X]) \cong K[X]$, so that the quotient field $q(\varphi(K[X]))$ of $\varphi(K[X])$ in L is isomorphic to $K(X)$. Since $q(\varphi(K[X]))$ contains $K \cup \{\theta\}$, we have $q(\varphi(K[X])) = L$, i.e. $L = K(x)$.

In case f is irreducible, the natural projection $\pi : K[X] \to K[X]/(f)$ maps K isomorphically onto $\pi(K)$ which we identify with K and we regard

$$K \subset K[X]/(f).$$

Then the isomorphism $\bar{\varphi} : K[X]/(f) \underset{\to}{\sim} \varphi(K[X])$ is a K-isomorphism (i.e. identity on K).

If $f(X) = \sum_{k=0}^{n} a_k X^k$, then putting $\pi(x) = \theta$, we obtain $f(\theta) = \sum_{k=0}^{n} a_k \pi(x)^k = \pi(f) = 0$ (in $K[X]/(f)$). Hence θ is a root of the polynomial f, completing the proof.   □

Definition 1.16. An element θ of an extension field L over K is called **algebraic** over K if there is a non-zero polynomial $f \in K[X]$ such that $f(\theta) = 0$. An extension field L over K is called **algebraic** over K if all the elements of L are algebraic over K. If $f(X)$ is an irreducible polynomial for θ in the sense of Theorem 1.14, it is of the least degree. Hence if we impose the monicness, i.e. the leading coefficient 1, then it is uniquely determined and is called the **minimal polynomial** of θ and denoted $\mathrm{Irr}(\theta, X, K)$. Its degree is called the **degree** of θ.

For a (not necessarily irreducible) polynomial in $K[X]$, an algebraic extension L of K is called a splitting field of f if f decomposes into the product $c(X - \theta_1) \cdots (X - \theta_1)$ of linear factors in $L[X]$ (c being the leading coefficient of f which we may assume to be 1). The field $K(\theta_1, \ldots, \theta_n)$, the smallest splitting field of f over K, is called the (minimal) **splitting field** of f over K.

Theorem 1.15. *An element θ of an extension L of K to be algebraic over K it is necessary and sufficient that*

$$[K(\theta) : K] < \infty \tag{1.30}$$

and if $\deg \operatorname{Irr}(\theta, X, K) = n$, then each element of $K(\theta)$ can be expressed uniquely as

$$a_0 + a_1\theta + \cdots + a_{n-1}\theta^{n-1}. \tag{1.31}$$

Proof. By Theorem 1.14, θ is algebraic if and only if $[K(\theta) : K] < \infty$ in which case $\operatorname{Irr}(\theta, X, K) = f(X)$ and any element of $K(\theta)$ is written in the form $g(\theta)$ with $g \in K[X]$. By the Euclidean division, there are polynomials $q, r \in K[X]$ such that

$$g = fq + r, \quad 0 \le \deg r < \deg f.$$

Hence $g(\theta) = r(\theta)$ and (1.31) follows. If there is another expression $b_0 + b_1\theta + \cdots + b_{n-1}\theta^{n-1}$, then we have $(a_0 - b_0) + (a_1 - b_1)\theta + \cdots + (a_{n-1} - b_{n-1})\theta^{n-1} = 0$ and since θ is of degree n, we must have $a_i = b_i$, and the expression (1.31) is unique. This completes the proof.   □

Theorem 1.16. *An extension of a field K obtained by adjoint of algebraic elements over K is algebraic.*

Proof. Let L be an extension of K and let $S \subset L$ be a set consisting of algebraic elements of L over K. By Proposition 1.14, $K(S) = \cup_F K(F)$, where F are finite sets. Hence it suffices to consider the case $L = K(\theta_1, \cdots, \theta_n)$ with θ_i being algebraic over K. Since θ_1 is algebraic over $K(\theta_1, \cdots, \theta_{i-1})$, it generates a finite extension over $K(\theta_1, \cdots, \theta_{i-1})$ by Theorem 1.15. Hence

$$[L : K] = [K(\theta_1, \cdots, \theta_n) : K(\theta_1, \cdots, \theta_{i-1})] \cdots [K(\theta_1) : K] < \infty.$$

Now if α is an arbitrary element of L, it generates an intermediate field $M = K(\alpha)$ between K and L, whence it is finite over K. Hence by Theorem 1.15, it is algebraic over K. Hence any element of L is algebraic over K and so L/K is algebraic, completing the proof.   □

Corollary 1.5. *A finite extension is a finitely generated algebraic extension.*

Proof. In the proof of Theorem 1.16 above it is proved that a finite extension is algebraic. Also, since the basis of L over K forms a finite set of generators of the extension, L/K is f.g. The converse is proved in the proof of Theorem 1.16. $\qquad\square$

Theorem 1.17. *Any non-zero polynomial over a field K has the splitting field, which is unique up to a K-isomorphism.*

Proof. If f has no prime factor of degree ≥ 2, then K itself is a splitting field. Suppose f has a prime factor and let $f_1(X)$ be the product of all prime factors of f of degree ≥ 2 and let $p(X)$ be one of them. We apply induction on the degree of f_1. By Theorem 1.14, it has a root in $L_1 := K[X]/(p)$. If we decompose f in L_1, then (p being decomposed) the degree of the product of irreducible factors of f (in $L_1[X]$) is less than $\deg f_1$. Hence induction applies. Uniqueness is omitted, completing the proof. $\qquad\square$

Remark 1.5. Any field K has its **algebraic closure** $\bar{K}$ by Steiniz' theorem, which is the field consisting of all algebraic elements of an (algebraically closed) extension over K. Here a field Ω is called an **algebraically closed field** if any (non-zero) polynomial in $\Omega[X]$ decomposes into linear factors. If we adjoin all the roots $\theta_1, \cdots, \theta_n \in \Omega$ of a polynomial $f \in K[X]$, then the splitting filed $K(\theta_1, \cdots, \theta_n)$ arises. Thus, *field theory is for decomposing polynomials to find their roots*. Since we consider extensions of the rational field, we may appeal to the algebraically closed field $\mathbb{C}$ and we work with $\bar{\mathbb{Q}}$, the algebraic closure of $\mathbb{Q}$.

Theorem 1.18. *Let L be an extension of K_1 and let Ω be an algebraic closure containing K_2. Then every isomorphism $\lambda : K_1 \to K_2$ is extended to an isomorphism $L \to \Omega$.*

Exercise 18. Let Ω be an algebraic closure of L. Then every isomorphism $L \to \Omega$ is extended to an automorphism of Ω.

Solution. In Theorem 1.18, choose Ω/L, $\Omega/\lambda(\Omega)$ as L/K_1, Ω/K_2 and $\lambda : L \to \lambda(L)$ be the isomorphism in the theorem. Then it can be extended to an isomorphism $\mu : \Omega \to \Omega$. Since $\mu(\Omega)$ is an algebraic closure of $\mu(L)$, which is $\lambda(L)$ and Ω is an algebraic extension of $\mu(\Omega)$, it follows that $\mu(\Omega) = \Omega$. Hence μ is an automorphism.

Corollary 1.6. *Let Ω be an algebraic closure of K and let $\alpha, \beta \in \Omega$. For the minimal polynomials of α, β coincide it is necessary and sufficient that there exists a K-automorphism λ of Ω such that $\alpha^\lambda = \beta$.*

Proof. Let $f(X) = \mathrm{Irr}(\alpha, K, X)$. If $\lambda(\alpha) = \lambda(\beta)$, then $f(\beta) = f(\alpha^\lambda) = f^\lambda(\alpha) = 0$. Hence $\mathrm{Irr}(\beta, K, X) = f(X)$.

Conversely, if $\mathrm{Irr}(\alpha, K, X) = \mathrm{Irr}(\beta, K, X)$, there exists a K-isomorphism $K(\alpha) \to K(\beta)$ such that $\alpha^\lambda = \beta$. By Exercise 18, this can be extended to a K-automorphism μ of Ω and $\alpha^\mu = \alpha^\lambda = \beta$. $\qquad\square$

Those two elements $\alpha, \beta \, (\in \Omega)$ whose minimal polynomials coincide are called **conjugate**. Among algebraic extensions (cf. Definition 1.15), those which arise from the irreducible polynomials without multiple roots are important. Suppose θ is an algebraic element of an extension L over K and that $\mathrm{Irr}(\theta, K, X)$ has no multiple roots (in $\bar{K}$), in which case both the polynomial and the element are called **separable**. An algebraic extension L/K is called **separable** if all the elements of L are separable over K.

Theorem 1.19. *A finite extension L/K is separable if and only if the number of distinct K-isomorphisms $L \to \bar{K}$ is equal to the extension degree.*

Separable extensions have many similar properties to algebraic extensions. The counterpart of Theorem 1.16 is true: *Adjoint of separable elements give rise to a separable extension.* The following two theorems will be of importance for us in Chapter 2.

Theorem 1.20. *A finite separable extension is a simple extension.*

Hence Theorem 1.15 is applicable.

Theorem 1.21. *If L/K is a finite separable extension, then there exists a separable, normal extension $L' \supset L$. If $[L : K] < \infty$, then L' can be taken of finite degree.*

Here a **normal extension** L/K means that if any irreducible polynomial $f \in K[X]$ has a root in L, then f decomposes into linear factors in $L[X]$, i.e. all the conjugate roots are in L. This may be stated that a normal extension is closed under $(K\text{-})$isomorphisms.

Theorem 1.22. *A finite extension L/K is normal if and only if it is a splitting field of a polynomial $K[X]$.*

Thus we may naively think of a normal extension as one for decomposing a polynomial (we can always do this if we go to $\bar{K}$, which is too big) and we are to choose the smallest, and a separable extension as one for a polynomial without multiple roots.

1.6 Applications to elementary number theory

We have **Fermat's little theorem**: For any $a \in \mathbb{Z}, (a, q) = 1$,

$$a^{\varphi(q)} \equiv 1 \,(\text{mod } q). \tag{1.32}$$

In particular, if p is a prime, then $\varphi(p) = p-1$, and for any $a \in \mathbb{Z}, (a, p) = 1$, we have

$$a^{p-1} \equiv 1 \,(\text{mod } p). \tag{1.33}$$

Indeed, the generalized Fermat's little theorem (1.23) coupled with (1.25) implies (1.32), and *a fortiori*, (1.33) follows.

Thus, many facts in elementary number theory can be clearly grasped through algebra.

Exercise 19. Let p be an odd prime. Then prove **Wilson's theorem**

$$(p-1)! \equiv -1 \,(\text{mod } p). \tag{1.34}$$

Deduce (1.33) from (1.34).

Solution. We may deduce (1.33) from (1.34) by rewriting (1.34) as $-\bar{1} = \bar{1} \cdot \bar{2} \cdots \overline{(p-1)}$ and multiplying by $\bar{n}^{p-1}$, we are done.

Proof of Wilson's theorem. We consider the polynomial $g(x) = x^{p-1} - \bar{1}$, which is a non-zero polynomial of degree $p-1$ over the field $K = \mathbb{Z}/p\mathbb{Z}$. By Fermat's little theorem, it has $(p-1)$ roots $\bar{1}, \bar{2}, \ldots, \overline{p-1}$. Hence $g(x) = (x - \bar{1})(x - \bar{2}) \cdots (x - \overline{p-1})$. Comparing the constant term, we conclude that $\bar{1} \cdot \bar{2} \cdot \cdots \cdot \overline{p-1} = (-1)^{p-1} \bar{1} \cdot \bar{2} \cdot \cdots \cdot \overline{p-1} = -\bar{1}$, which is the same as (1.34).

Another proof. We note that $\bar{x}^2 = \bar{1}$ has only two solution $\bar{x} = \pm\bar{1}$. Other $\frac{p-3}{2}$ elements have their respective $\frac{p-3}{2}$ inverses. Hence $\prod_{i=1}^{p-1} i = \bar{1} \cdot \overline{p-1} \cdot \prod_{i=2}^{p-3} i = -\bar{1}$ and the result follows.

1.7   Chinese remainder theorem

Definition 1.17. We have tacitly assumed in the **direct product** of algebraic systems, the componentwise operation is defined to make it an algebraic system. The direct product (or sometimes called **direct sum** if the number of summands is finite) is an effective way of constructing new algebraic systems, along with forming quotient groups and tensor products. All operations in direct products are assumed to be done **componentwise**. For example, in the direct sum of groups $G_1 \times \cdots \times G_n$, the operation is to be $(a_1, \cdots, a_n)(b_1, \cdots, b_n) = (a_1 b_1, \cdots, a_n b_n)$. The same holds true for other operations.

The contents of this section is an extract from [Vis2, §B6, pp.241-245]. Theorem 1.23 is fundamental and there are many proofs, all of which depend on the following fact. *If $m_1, m_2, \cdots, m_n$ are positive integers, pairwise prime, then $x \equiv y \pmod{m_i}$, $1 \leq i \leq n$ if and only if $x \equiv y \pmod{m_1 \cdots m_n}$.*

Theorem 1.23. *Let $m_1, m_2, \cdots, m_n$ be positive integers, pairwise prime, $(m_i, m_j) = 1$, $i \neq j$. Then for any integers $c_1, c_2, \cdots, c_n$, there exists a unique residue x mod $m_1 \cdots m_n$ such that*

$$x \equiv c_i \pmod{m_i}, \ 1 \leq i \leq n.$$

Proof. Let $\mathbb{Z}/m_1\mathbb{Z} \times \cdots \times \mathbb{Z}/m_n\mathbb{Z}$ indicate the Cartesian product of the residue class rings mod m_i, $1 \leq i \leq n$ and let f be the ring homomorphism

$$f : \mathbb{Z} \longrightarrow (\mathbb{Z}/m_1\mathbb{Z}) \times \cdots \times (\mathbb{Z}/m_n\mathbb{Z}),$$

defined by

$$f(a) = (a + m_1\mathbb{Z}, \ \cdots, \ a + m_n\mathbb{Z}), \ a \in \mathbb{Z}.$$

Since Ker $f = \{a \in \mathbb{Z} \mid a \equiv 0 \pmod{m_i}, \ 1 \leq i \leq n\}$, it follows that

$$\mathrm{Ker} f = \{a \in \mathbb{Z} \mid a \equiv 0 \pmod{m_1 \cdots m_n}\} = m_1 \cdots m_n\mathbb{Z}.$$

Hence, by the ring homomorphism theorem (Theorem 1.12), we obtain the isomorphism

$$\mathbb{Z}/m_1 \cdots m_n\mathbb{Z} \cong \mathrm{Im} f \subset (\mathbb{Z}/m_1\mathbb{Z}) \times \cdots \times (\mathbb{Z}/m_n\mathbb{Z}).$$

The cardinality of $\mathbb{Z}/m_1 \cdots m_n\mathbb{Z}$ and that of $(\mathbb{Z}/m_1\mathbb{Z}) \times \cdots \times (\mathbb{Z}/m_n\mathbb{Z})$ is $m_1 \cdots m_n$, and therefore f is surjective, whence we have the **direct sum decomposition**

$$\tilde{f} : \mathbb{Z}/m_1 \cdots m_n\mathbb{Z} \cong (\mathbb{Z}/m_1\mathbb{Z}) \times \cdots \times (\mathbb{Z}/m_n\mathbb{Z}), \qquad (1.35)$$

with

$$\tilde{f}(a + m_1 \cdots m_n \mathbb{Z}) = (a + m_1 \mathbb{Z}, \ \cdots, \ a + m_n \mathbb{Z}),$$

Hence for any $c_i + m_i \mathbb{Z}, 1 \le i \le n$, there is an $x + m_1 \cdots m_n \mathbb{Z}$ such that $x \equiv c_i \pmod{m_i}, 1 \le i \le n$, which is Theorem 1.23. $\qquad\square$

Remark 1.6. Mozzochi [Mozz] gave a proof of Theorem 1.23 by providing the $\tilde{f}$ in (1.35).

From (1.35) it follows that a unit is mapped onto a unit, whence they are equal in number:

$$|(\mathbb{Z}/m_1 \cdots m_n \mathbb{Z})^\times| = |(\mathbb{Z}/m_1 \mathbb{Z})^\times| \times \cdots \times |(\mathbb{Z}/m_n \mathbb{Z})^\times|,$$

where as in Definition 1.3, $R^\times$ indicates the unit group of the ring R.

Now, by (1.25), the order of the unit group of $\mathbb{Z}/m\mathbb{Z}$ being the Euler function $\varphi(m)$, we conclude that

$$\varphi(m_1 \cdots m_n) = \varphi(m_1) \cdots \varphi(m_n).$$

In particular, **multiplicativity** of the Euler function follows

$$\varphi(mn) = \varphi(m)\varphi(n), \ \gcd(m,n) = 1. \tag{1.36}$$

Noting that for a prime power, we have

$$\varphi(p^\alpha) = p^\alpha - p^{\alpha-1} = p^\alpha \left(1 - \frac{1}{p}\right), \tag{1.37}$$

we may deduce an explicit formula for φ:

$$\varphi(m) = m \prod_{p|m} \left(1 - \frac{1}{p}\right). \tag{1.38}$$

Introducing the **Möbius function**

$$\mu(n) = \begin{cases} (-1)^s, & n = p_1 \cdots p_s, \quad p_i \quad \text{distinct primes} \\ 1, & n = 1, \\ 0, & p^2 | n, \end{cases} \tag{1.39}$$

we may express (1.38) in a more familiar manner:

Exercise 20. From (1.38), prove the identity

$$\varphi(m) = m \sum_{d|m} \frac{\mu(d)}{d}. \tag{1.40}$$

Also prove that (1.40) and (1.27) are reciprocal relations under (1.41), which is an example of the **Möbius inversion** (Corollary 3.4).

Exercise 21. Let $\zeta(s)$ denote the Riemann zeta-function to be introduced in (3.69) in Chapter 3. Then prove that

$$\sum_{n=1}^{\infty} \frac{\mu(n)}{n^s} = \frac{1}{\zeta(s)}$$

for $\operatorname{Re} s := \sigma > 1$ and whence (3.3) in Chapter 3, to the effect that

$$\sum_{d|n} \mu(d) = \begin{cases} 0, & n > 1 \\ 1, & n = 1. \end{cases} \tag{1.41}$$

Also deduce (1.41) from (1.39).

Another Proof of Theorem 1.23. Let $m = m_1 \cdots m_n$. Then since $(m_i, m/m_i) = 1$, by Theorem 1.6, there exist integers a_i, b_i such that $a_i m_i + b_i m/m_i = 1$. Putting $M_i = 1 - a_i m_i = b_i m/m_i$, we see that

$$M_i \equiv \begin{cases} 1 \ (\text{mod } m_i); \\ 0 \ (\text{mod } m_j), \ j \neq i. \end{cases}$$

Then $x = \sum_{i=1}^{n} c_i M_i$ is the required solution.

If there are two solutions x, $y \mod m$, i.e. if $x \equiv c_i$, $y \equiv c_i \ (\text{mod } m_i)$, then $x \equiv y \mod m_i$, whence $x \equiv y \mod m$ follows.

Exercise 22. Let

$$A = \{x \mid 0 \leq x \leq m_1 \cdots m_n - 1\}$$

be the set of least non-negative residues mod $m_1 \cdots m_n$ and

$$B = \{(x_1, \cdots, x_n) \mid 0 \leq x_i \leq m_i - 1, \ 1 \leq i \leq n\}$$

be the Cartesian product of the sets of least non-negative residues mod $m_i, 1 \leq i \leq n$, both consisting of $m_1 \cdots m_n$ elements. For each $x \in A$ choose x_i, $0 \leq x_i \leq m_i - 1$ such that

$$x_i \equiv x \ (\text{mod } m_i), \ 1 \leq i \leq n$$

and define the map

$$f : A \to B; \ f(x) = (x_1, \cdots, x_n)$$

and prove Theorem 1.23.

For a solution, cf. [Vis2, pp. 243-244].

The name of "Sunzi" is well-known to many of the Chinese and the Japanese by the passage from his War strategy book.

The origin of the Chinese remainder theorem lies in

$$N \equiv 2 \pmod 3, \equiv 3 \pmod 5, \equiv 2 \pmod 7.$$

The least positive solution to this system of congruences is $N = 23$. Cf. [L.-K. Hua, Introduction to number theory, pp. 32-33].

Corollary 1.7. *Suppose* $m_1, \cdots, m_n$ *are pairwise prime,* $(m_i, m_j) = 1, i \neq j$ *and that* $(a_i, m_i) = 1$, $1 \le i \le n$. *Then the system of congruences*

$$a_i x \equiv c_i \pmod{m_i}, 1 \le i \le n,$$

has a unique solution mod $m_1 \cdots m_n$.

The equation being the same as

$$x \equiv a_i^{-1} c_i \pmod{m_i}, \ 1 \le i \le n,$$

Corollary 1.7 is apparent from the Chinese remainder theorem. For another proof, cf. [Vis2, pp. 244-245].

1.8 Applications of the gcd principle

Theorem 1.24. *Let* $f(n)$ *be an arithmetical function, periodic of period* q:

$$f(n + q) = f(n), \qquad n \in X \subset \mathbb{Z},$$

where we suppose for X *all the subsequent operations are valid. Then the least period* f *must divide other periods.*

Proof. We prove the g.c.d. of two periods q_1, q_2 is again a period. Since by Theorem 1.6,

$$(q_1, q_2) = d = aq_1 + bq_2, \qquad a, b \in \mathbb{Z},$$

it follows that

$$f(n + d) = f(n + aq_1 + bq_2) = f(n + bq_2) = f(n).$$

Hence the gcd principle (Theorem 1.4) applies. $\qquad \square$

Exercise 23. Give another proof of Proposition 1.3 to the effect that any subgroup of a cyclic group is again cyclic.

Solution.   We may prove this as a simple manifestation of the gcd principle with $f(m)$ being such that $a^{f(m)} \in H$.

We recall that the set $\widehat{(\mathbb{Z}/q\mathbb{Z})}^{\times}$ of all residue class characters χ mod q forms a (multiplicative) group of order $\varphi(q)$ (being the Euler function), with the principal character χ_0^* having all values 1 as its identity and $\chi^{-1} = \bar{\chi}$. For this cf. Theorem 4.1 in Chapter 4.

The generating function of the above-mentioned conventional principal (primitive) character $\chi_0^*(n) = 1$ for all n ([Prac, p. 127]) is the Riemann zeta-function: $\zeta(s) = \sum_{n=1}^{\infty} \frac{\chi_0^*(n)}{n^s}$ for $\sigma = \mathrm{Re}\, s > 1$.

That it is isomorphic to the group $(\mathbb{Z}/q\mathbb{Z})^{\times}$ of reduced residue classes mod q, cf. Theorem 4.3 in Chapter 4. For some more details, cf. Chapter 3, §3.8.

Definition 1.18. We define an arithmetical function $\tilde{\chi}$ from χ mod q as follows. For $(a, q) = 1$, we let $\tilde{\chi}(a) = \chi(\bar{a})$, with $\bar{a}$ indicating the reduced residue class containing a. Then we call its 0-extension (i.e. $\tilde{\chi}(a) = 0$ if $(a, q) > 1$) a Dirichlet character $\tilde{\chi}$ mod q, and we use the same symbol χ to indicate this arithmetical function and refer to it as a **Dirichlet character** χ mod q. They inherit the **multiplicative** property $\chi(ab) = \chi(a)\chi(b)$ of residue class characters and the **principal Dirichlet character**, denoted by χ_0, is a periodic function of period q having the values $\chi_0(a) = 1$ or $\chi_0(a) = 0$ according as $(a, q) = 1$ or $(a, q) > 1$. A divisor q_1 of q is called a **defining** (or **admissible**) **modulus** for χ if $(a, q) = 1$, $a \equiv 1 \pmod{q_1}$ imply $\chi(a) = 1$, which amounts to saying that $\chi(a) = \chi(b)$ if $a \equiv b \pmod{q_1}$, $(a, q) = (b, q) = 1$. This last statement is the same as χ being periodic of period q_1.

If $q_1, q_2 | q$ are defining moduli of $\tilde{\chi}$, then so is (q_1, q_2) and the principle implies that the least defining modulus $f = f_{\chi}$ (called the **conductor** of $\tilde{\chi}$, f being from the German word "Fuerer" meaning a tram conductor) divides all other defining moduli, and in particular, (q_1, q_2). If the conductor f is q, then χ as well as $\tilde{\chi}$ is called a **primitive** character to the modulus f, or mod f (read: modulo f), otherwise **imprimitive**, i.e. if there is a proper divisor f of q such that f is a defining modulus of χ as well as $\tilde{\chi}$, or in other words, χ is **induced** from ψ through the canonical surjection $(\mathbb{Z}/q\mathbb{Z})^{\times} \to (\mathbb{Z}/f\mathbb{Z})^{\times}$. In this case, χ is identical with a primitive character

ψ to the modulus f, where ψ is defined as follows. If $(n, q) = 1$, then $\psi(n) = \chi(n)$, and if $(n, q) > 1$ but $(n, f) = 1$, then we choose an integer a satisfying $(n + af, q) = 1$ and define $\psi(n) = \chi(n + af)$. We need to prove that this ψ is one of $\varphi(f)$ Dirichlet characters. For this we use the following characterization of Dirichlet characters, i.e. *An arithmetical function* X *is a Dirichlet character* mod q *if and only if it satisfies the following conditions: multiplicative, periodic of period* q, $\mathrm{X}(n) = 0$ *for* $(n, q) > 1$ *but not always* 0.

Exercise 24. Prove the above characterization of Dirichlet characters.

Proof. We apply the **orthogonality** relation (cf. Proposition 4.3)

$$\sum_{\chi} \chi(n) = \begin{cases} \varphi(q), & \text{if } n \equiv 1 \,(\text{mod } q) \\ 0, & \text{otherwise.} \end{cases}$$

For $(c, q) = 1$, we have

$$\sum_{n(\text{mod } q)} \bar{\chi}\mathrm{X}(n) = \sum_{n(\text{mod } q)} \bar{\chi}\mathrm{X}(cn) = \bar{\chi}\mathrm{X}(c) \sum_{n(\text{mod } q)} \bar{\chi}\mathrm{X}(n),$$

whence we have either $\mathrm{X} = \chi$ or $\sum_{n(\text{mod } q)} \bar{\chi}\mathrm{X}(n) = 0$. In the latter case, we have for all m

$$0 = \sum_{\chi} \chi(m) \sum_{n(\text{mod } q)} \bar{\chi}\mathrm{X}(n).$$

Changing the order of summation, we transform the right-hand side into

$$\sum_{n(\text{mod } q)} \mathrm{X}(n) \sum_{\chi} \chi(m)\bar{\chi}(n),$$

which leads to $\mathrm{X}(m) = 0$ for all m contrary to the hypothesis. □

By the above remark, we see that to each imprimitive character χ mod q, there corresponds a (unique) primitive character ψ mod f, f being the conductor of ψ such that

$$\chi(n) = \chi_0\psi(n) = \begin{cases} \psi(n), & \text{if } (n, q) = 1 \\ 0, & \text{if } (n, q) > 1, \end{cases} \tag{1.42}$$

where χ_0 is the principal Dirichlet character induced by the principal residue class character χ_0^*. If χ is related to ψ by (1.42), then we say that ψ **induces** χ. It is clear that χ defined for any multiple q of f by (1.42) is a Dirichlet character mod q.

(1.42) is equivalent to

$$L(s, \chi) = L(s, \psi) \prod_{p|\bar{q}} \left(1 - \psi(p)p^{-s}\right), \tag{1.43}$$

where $L(s, \chi)$ indicates the Dirichlet L-function defined by (5.3), Chapter 5. For a proof, cf. Exercise 58 in Chapter 3. The above argument now establishes the following principle corresponding to the one stated in §1.

In the character group $\hat{G} = \widehat{(\mathbb{Z}/q\mathbb{Z})}^{\times}$, for each divisor d of q, there are $f(d)$ primitive characters $\bmod\, d$ and this gives a classification of elements of $\hat{G}$ into disjoint subsets.

In particular, the cardinality relation is

$$\varphi(q) = \sum_{d|q} f(d). \tag{1.44}$$

By Corollary 3.4, we may invert (1.44) to obtain ([BShh, p. 420])

Corollary 1.8. *The number $f(q)$ of primitive characters $\bmod\, q$ is given by*

$$f(q) = \sum_{d|q} \mu(d)\varphi\left(\frac{q}{d}\right). \tag{1.45}$$

(1.44) is to be compared with the well-known important fact ([BShh, p. 388]) that

$$\zeta_k(s) = \zeta(s) \prod_{\substack{\chi \text{ primitive mod } d \\ 1 \neq d | m}} L(s, \chi), \tag{1.46}$$

where $\zeta_k(s)$ indicates the Dedekind zeta-function of the m-th cyclotomic field k (cf. Chapter 2).

1.9 Group actions

Definition 1.19. Let $X \neq \phi$ be a set and let G be a group. We say that G **acts on** X if for any $a \in G, x \in X$, we have $ax \in X$ and
(i) $a, b \in G$, $x \in X$, $(ab)(x) = a(bx)$
and

(ii) $1x = x$

are satisfied. We also say that G is a **transformation group** on X, and denote it by (G, X) or simply by G. If we introduce the relation among elements $x, y \in X$ in the following way, then this yields a classification of X: $x \sim y \Leftrightarrow$ there is an $a \in G$ such that $ax = y$.

Exercise 25. Prove that this is an equivalence relation.

Exercise 26. Let X be a finite set, $\{1, 2, \cdots, n\}$, say and let $S(X)$ or $\mathrm{Sym}(X)$ denote the set of all bijections on X. Then prove that $S(X)$ is a group acting on X, called the **symmetric group** of order n and also denoted S_n. Prove that each element can be decomposed into a product of transitions (i.e. permutes only 2 elements), with the predetermined parity of the number of transitions. Then we can define the signature $\mathrm{sgn}\,\sigma$ of each permutation $\sigma \in S_n$ by $(-1)^{\mathrm{sgn}\,\sigma} = 1$ if σ consists of an even number of transitions and -1 if σ consists of an odd number of transitions. Prove that $S_n \to \{\pm 1\}$ is a homomorphism.

Exercise 27. ([DrSh]) Let X_i, $1 = 1, \cdots, n$ be finite sets and let $X = X_1 \times \cdots \times X_n$ be their Cartesian product and let $y_i = \sharp X / \sharp X_i$. Let $\sigma_i \in S(X_i)$ and let $\sigma = (\sigma_1, \cdots, \sigma_n) \in S(X)$ be the permutation obtained by applying each σ_i to X_i. Then prove that

$$\mathrm{sgn}\,\sigma = \prod_{i=1}^{n} \mathrm{sgn}\,\sigma_i{}^{y_i}. \tag{1.47}$$

Solution. For each i, let $\hat{\sigma}_i$ be an extension of σ_i, i.e. it is obtained by applying σ_i to the i-th coordinate and the identity to other coordinates. Then $\hat{\sigma}_i$ is the y_i copies of σ_i, it follows that $\mathrm{sgn}\,\hat{\sigma}_i = (\mathrm{sgn}\,\sigma_i)^{y_i}$ and that

$$\sigma = \hat{\sigma}_1 \cdots \hat{\sigma}_n.$$

Hence (1.47) follows.

The class containing $x \in X$, written C_x, is called an **orbit** (or a **transitive domain**) of x and we denote the set of all orbits by $G \backslash X$: $G \backslash X = \{C_x | C_x \cap C_y = \phi\}$. If $\sharp G \backslash X = 1$, i.e. for every element $y \in X$, there is a $\sigma \in G$ and $x \in X$ such that $y = ax$, then G is said to act on X transitively or G is a transitive group on X.

Theorem 1.25. (Burnside) *Let G be a transformation group on X. Then for a fixed element $x \in X$, the set $\mathcal{F}(x) = \{a \in G | ax = x\}$ forms a subgroup of G (called the **stabilizer** (of x)) and we have $|G| = \sharp C_x |\mathcal{F}(x)|$, or*

$$\sharp C_x = (G : \mathcal{F}(x)). \tag{1.48}$$

Proof. It is easy to see that $\mathcal{F}(x)$ forms a subgroup. Also it is easy to prove that the relation $a \sim b$ defined in G as follows, is an equivalence relation: $a \sim b$ if $ax = bx$. We denote the equivalence class containing a by $\widetilde{C_a}$. Since each element ax of C_x gives rise to one $\widetilde{C_a}$ of G consisting of all elements b of G such that $bx = ax$, G is decomposed into $\sharp C_x$ equivalence classes: $G = \overset{\sharp C_x}{\underset{i=1}{\cup}} \widetilde{C_{a_i}}$ (disjoint).

Now, since $b \in \widetilde{C_a}$ is equivalent to $bx = ax$, or $a^{-1}b \in H_x$, or to $b \in a\mathcal{F}(x)$, we have $\#\widetilde{C_a} = \#a\mathcal{F}(x) = |\mathcal{F}(x)|$. Hence each class $\widetilde{C_{a_i}}$ in $G/\sim$ contains the same number $|\mathcal{F}(x)|$ of elements, and so this proves the assertion. □

Example 1.5. Recall Example 1.3 in which we defined the equivalence of two elements a, b of a group G modulo a subgroup H of G. We may interpret Example 1.3 as the action of a subgroup H on $X = G$. Indeed, defining the action of H on G by $a(x) = ax$, $a \in H$, $x \in G$, we find that the orbit containing $x \in G$ is Hx, the right coset containing x, whose cardinality is $|H|$ and the stabilizer of x is 1. Hence (1.48) leads to a triviality $|H| = |Hx|$. But compare (1.22).

Example 1.6. If G is a group, two elements $x, y \in G$ are called **conjugate** if there is an element $a \in G$ such that $y = a^{-1}xa$. Viewed as a mapping, the correspondence $I_a(x) = a^{-1}xa$ is called the **inner automorphism**. Let

$$C_G(x) = \{a \in G | a^{-1}xa = x\} \tag{1.49}$$

be the **centralizer** of $x \in G$. We may define an action of G on G by

$$a(x) = I_a(x) = a^{-1}xa. \tag{1.50}$$

Then the orbit containing x is called the **conjugate class** containing x and the centralizer is the stabilizer $\mathcal{F}(x)$ of x. Hence Theorem 1.25 implies that the number g_x of elements of the conjugate class C_x containing x has $(G : C_G(x))$ elements and *a fortiori* is a divisor of $|G|$. Since the conjugate class decomposition is a classification of G, we have the **class equation**

$$|G| = \sum_{\substack{x \in G \\ x: \text{non-conjugate}}} g_x. \tag{1.51}$$

Exercise 28. Prove that the conjugacy is an equivalence relation, that (1.50) defines an action and that the centralizer is a subgroup. Also prove that the inner automorphism is an automorphism.

Example 1.7. In the same spirit as in Example 1.6 define the action of a group G on the set X of all its subgroups: $X = \{H \subset G | H$ is a subgroup of $G\}$ by

$$a(H) = I_a(H) = a^{-1}Ha = \{a^{-1}xa | x \in H\}. \tag{1.52}$$

Similarly to (1.53) we define the **normalizer** $N_G(H)$ of H by

$$N_G(H) = \{a \in G | a^{-1}Ha = H\}. \tag{1.53}$$

Then the stabilizer of H is the normalizer of H and the number of conjugate subgroups of H is given by $(G : N_H(x))$.

Exercise 29. Prove that the conjugacy of subgroups is an equivalence relation, that (1.7) defines an action and that the normalizer is a subgroup. Also prove that $a^{-1}Ha$ is isomorphic to H.

Exercise 30. Interpret the additive group $\mathbb{Z}/q\mathbb{Z}$ consisting of residue classes modulo q in Example 1.4 in the context of Example 1.5.

Solution. Indeed, the action of $q\mathbb{Z}$ on $\mathbb{Z}$ is defined by $a(x) = a + x$, $a \in q\mathbb{Z}$, $x \in \mathbb{Z}$ and we have the quotient group Example 1.4 whose order is q.

Exercise 31. (i) Let G be a transitive subgroup of $S(X)$. Then prove that $\sharp G = \sharp X$ and $\sigma \in G$ is a square in G if and only if $\operatorname{sgn} \sigma = 1$.
(ii) Let $N \lhd G$ and let $\sharp N$ is odd. Then prove that $xN \in G/N$ is a square if and only if x is a square in G.

Solution. Use Theorem 1.25.

Exercise 32. (i) Let $\mathcal{H}$ denote the upper half-plane $\mathcal{H} = \{z \in \mathbb{C} | \operatorname{Im} z > 0\}$ and let $\mathrm{SL}(2, \mathbb{Z}) = \left\{ \gamma = \begin{pmatrix} a & b \\ c & d \end{pmatrix} \middle| ad - bc = 1 \right\}$ the full modular group.

Define the action of $\gamma = \begin{pmatrix} a & b \\ c & d \end{pmatrix} \in \mathrm{SL}(2, \mathbb{Z})$ on $\mathcal{H}$ by

$$\gamma z = \frac{az + b}{cz + d}. \tag{1.54}$$

Prove that $\operatorname{Im} \gamma z = \frac{\operatorname{Im} z}{|cz+d|^2}$, so that the above definition makes sense.
(ii) Denote the set of all rational functions γ defined above by (1.54) by G. Then prove that

$$G \cong \mathrm{SL}(2, \mathbb{Z})/\{\pm I\},$$

the latter of which is denoted by $\mathrm{PSL}(2, \mathbb{Z})$, where I is the identity matrix.

Solution.   The multiplication in G is defined by the composition $\gamma_1 \circ \gamma_2(z) = \gamma_1(\gamma_2(z))$, which makes G a group and it can be checked that $\gamma_1 \circ \gamma_2(z) = (\gamma_1\gamma_2)(z)$, where on the right-hand side the product is the ordinary one for matrices. Hence defining a map $\varphi : \mathrm{SL}(2, \mathbb{Z}) \to G$ by $\varphi(\gamma) = \gamma$, we see that φ is a homomorphism.

Then $\mathrm{Ker}\,\varphi$ consists of those rational functions for which $\frac{az+b}{cz+d} = z$. Hence $c = 0$ and $a/d = 1$, and $b = 0$. Substituting into $\det \gamma = 1$, we see that the matrices must be $\pm I$. Now Theorem 1.10 applies.

$\mathrm{PSL}(2, \mathbb{Z})$ acts on $\mathcal{H}$ faithfully. Cf. §4.3 and Chapter 6 for further examples.

We note an interesting interpretation of the modular group as a free group in [Kuro, pp. 16-17; pp. 261-264], where the action is understood to be the concatenation.

1.10   Finite fields

Let F be a finite field (commutative or not) with the identity 1. The characteristic, $\mathrm{char}\,F$, is a prime number $p > 1$ and the prime ring in F is isomorphic to $\mathbb{F}_p = \mathbb{Z}/p\mathbb{Z}$ (cf. §1.5 above, especially Definition 1.13 and Corollary 1.4). F may be regarded as a vector space over $\mathbb{F}_p$ of finite dimension, say f. Then $\sharp F = p^f = q$, say. If F is a subfield of a finite field F' with $q' = p^{f'}$ elements, then F' may be regarded as a (left) vector space over F with dimension d, say. Then $f' = df$ and $q' = q^d = p^{df}$.

Our aim is this section is to prove the following theorems on finite fields.

Theorem 1.26. (Wedderburn) *All finite fields are commutative.*

Proof.   Let F be a finite field of $\mathrm{char}\,F = p$ and let Z be its center with $q = p^f$ elements. The center being a finite field, F may be regarded as a vector space over Z and if the dimension is d, then F has q^d elements. We want to prove $d = 1$. Applying Example 1.6 to the unit group $F^\times = F - \{0\}$, we see that the centralizer $N(x) = C_{F^\times}(x)$ of $x \in F^\times$ is a subfield of F containing Z. Let $\delta(x)$ be the dimension of $N(x)$ over Z. Then $\sharp N(x) = q^{\delta(x)}$ and as we have seen above, d is a multiple of $\delta(x)$ and $\delta(x) < d$ if $x \notin Z$. By Example 1.6, the number of elements of the conjugacy class C_x containing x is $(F^\times : N(x)) = \frac{q^d-1}{q^{\delta(x)}-1}$, and so the class equation (1.51) reads

$$q^d - 1 = q - 1 + \sum_x \frac{q^d - 1}{q^{\delta(x)} - 1}, \tag{1.55}$$

the sum being taken over the complete set of representatives of the non-central conjugacy classes of $F^\times$.

Now suppose that $d > 1$ and consider the d-th cyclotomic polynomial $\Phi_d(X) = \prod(X - \zeta)$, the product being taken over all primitive d-th roots of 1 in $\mathbb{C}$ (to be introduced in (2.29) in Chapter 2). We can prove that $\Phi_d \in \mathbb{Z}[X]$ (cf. remark after Theorem 2.1, Chapter 2). Clearly $\Phi_d \mid \frac{X^d-1}{X^{\delta(x)}-1}$ for $x \notin Z$. Hence all the terms on the left of (1.55) save for $q-1$ is divisible by $\Phi_d(q)$, so that $q-1$ must also be divisible by $\Phi_d(q)$. However, each factor of $\Phi_d(q) = \prod(q - \zeta)$ has absolute value $> q - 1$. This is a contradiction, so that we must have $d = 1$, completing the proof. $\square$

Hence hereafter, whenever we refer to a finite field, we mean a commutative finite field.

Theorem 1.27. *Every finite subgroup of the unit group of a commutative field is cyclic, and a fortiori the unit group of a finite field is cyclic.*

To prove this theorem we need a lemma.

Lemma 1.1. *Let G be an Abelian group of order n with the property that for all divisors d of n, there are at most d roots of the equation*

$$X^d = 1. \tag{1.56}$$

Then G is cyclic.

Proof. [Weil, p. 3] Let α be an element of G of maximal order N and let β be any element of G of order n. If n does not divide N, there is a prime p such that

$$q := p - \nu|n, \ N = N'p^\mu, \ (N',p^\mu) = 1, \ \mu < \nu.$$

We have $o(\alpha^{p^\mu}) = N'$ and $o(\beta^{\frac{n}{q}}) = q = p^\nu$. Since $(N',p^\nu) = 1$, it follows that

$$o(\alpha^{p^\mu}\beta^{\frac{n}{q}}) = N'p^\nu = Np^{\nu-\mu} > N,$$

a contradiction. Hence $n|N$.

Now since β is a root of $X^n - 1$ which has n distinct roots $\alpha^{\frac{kN}{n}}$, $0 \le k < n$, it follows that β is a power of α and α generates G. $\square$

Second proof. We use introduction on n and suppose the assertion is true for all groups with order $< n$. Then all proper subgroups of G, being of smaller than n, are cyclic.

First suppose that the canonical decomposition of n is such that $n = p_1{}^{e_1} \cdots p_r{}^{e_r}$ with $r \geq 2$. Then letting P_i denote the subgroup of G consisting of all elements whose orders are powers of $p_i, (1 \leq i \leq r)$, we have

$$G = P_1 \oplus \cdots \oplus P_r.$$

Since each P_i is cyclic and the orders are relatively prime, it follows that G is cyclic. (This a step for the proof of Chapter 4, Lemma 4.3.)

Suppose now $n = p^\epsilon$ and G is not cyclic. Then all elements of G satisfy the equation

$$a^{p^{\epsilon-1}} = 1,$$

which are p^ϵ in number. This contradicts the assumption with $d = p^{\epsilon-1}$.

Third proof. [Serr, p. 4] We shall show that for each $d|n$, the number of elements of G of order d is $\varphi(d)$. If there is an element a of G of order d, then all the elements of the subgroup $< a > = \{1, a, \cdots, a^{d-1}\}$ generated by a satisfy (1.56). Since $| < a > | = d$, by assumption, they exhaust all the solutions of (1.56) and so any element b of G such that $b^d = 1$ belongs to $< a >$. Hence, in particular, all the elements of G of order d are generators of $< a >$. Since it is a cyclic group, by Proposition 1.3 there are $\varphi(d)$ generators. If for some value of $d|n$, the number of elements of G of order d is $< \varphi(d)$, then it contradicts the equality (1.27). Hence, for all divisors d of n, there are $\varphi(d)$ elements of order d. and, in particular, there is an element of G of order n, i.e. G is cyclic.

Theorem 1.28. *(i) For any prime power $q = p^e$, there exists a unique finite field $\mathbb{F}_q$ with q elements, which is a splitting field of $X^q - X$ and $\mathbb{F}_q^\times$ is a cyclic group of order $q - 1$.*
(ii) For any $e \in \mathbb{N}$, there exists a unique extension $\mathbb{F}_{q^e}$ of $\mathbb{F}_q$ of degree e. The extension $\mathbb{F}_{q^e}/\mathbb{F}_q$ is a cyclic extension and the Galois group is generated by σ such that $\alpha^\sigma = \alpha^q$ $(\forall \alpha \in \mathbb{F}_{q^e})$.

Example 1.8. The unit group (reduced residue classes modulo p) of the prime field $\mathbb{Z}/p\mathbb{Z}$, where p is a prime number, is cyclic. Its generators g are called **primitive root**s modulo p:

$$(\mathbb{Z}/p\mathbb{Z})^\times = < g > .$$

Exercise 33. (Proof of the existence of a primitive root of 1) Let $F = \mathrm{GF}(q)$ and let N be relatively prime to p. Then there exists a primitive N-th root ζ in the extension field $E = GF(q^{\varphi(N)})$ of F.

Solution. Letting α be a primitive element of $E = GF(q^{\varphi(N)})$, we see that α has order $q^{\varphi(N)} - 1$ in $E^\times$. Since $q^{\varphi(N)} \equiv 1 \mod N$, we may put $\beta = \alpha^{\frac{\varphi(N)-1}{N}}$ and conclude that β is a primitive N-th root of 1 in E.

For an odd prime p, we define the **Legendre symbol** $\left(\frac{a}{p}\right)$ for $(a, p) = 1$ to be 1 or -1 according to a as a **quadratic residue** or **quadratic non-residue**, i.e. according to $b \in \mathbb{Z}$ such that $a \equiv b^2 \pmod{p}$ or not. We define $\left(\frac{a}{p}\right) = 0$ for $(a, p) > 1$, Then $\left(\frac{a}{p}\right)$ is a Dirichlet character in the sense of Definition 1.18. We have several useful properties of the Legendre symbols.

(1) (Euler's criterion) $\quad \left(\frac{a}{p}\right) \equiv a^{\frac{p-1}{2}} \pmod{p}$.

(2) (the first supplementary law) $\quad \left(\frac{-1}{p}\right) \equiv (-1)^{\frac{p-1}{2}} \pmod{p}$.

(3) (the second supplementary law) $\quad \left(\frac{2}{p}\right) \equiv (-1)^{\frac{p^2-1}{8}} \pmod{p}$.

The Euler criterion follows from the fact that $\mathbb{F}_p^\times$ is cyclic which implies the first supplementary law. The latter also follows by remarking that -1 is a square in $\mathbb{F}_p$ if and only there is a primitive 4-th root of unity.

Example 1.9. (Theorem 5.6) For distinct odd primes p, q, we have

$$\left(\frac{p}{q}\right)\left(\frac{q}{p}\right) = (-1)^{\frac{p-1}{2}}(-1)^{\frac{q-1}{2}}. \tag{1.57}$$

Proof. We present the neat proof of F. Keune [Keun], other proofs being given in later chapters. Let n be the order of the cyclic group $< \bar{p} > \subset \mathbb{F}_q^\times$. Then $p^n \equiv 1 \mod q$ and q is a divisor of the order $p^n - 1$ of $\mathbb{F}_{p^n-1}^\times$. By Theorem 1.28 and Proposition 1.6, there exists a generator of a subgroup of $\mathbb{F}_{p^n-1}^\times$ of order q, which turns out to be the same as a primitive q-th root of unity, ζ, say.

Now consider the matrix

$$A = \begin{pmatrix} 1 & 1 & \cdots & 1 \\ 1 & \zeta & \cdots & \zeta^{q-1} \\ & & \cdots & \\ 1 & \zeta^{q-1} & \cdots & \zeta^{(q-1)(q-1)} \end{pmatrix} \tag{1.58}$$

and let δ denote its determinant (the Vandermonde determinant). Then δ^2 is the discriminant of the polynomial $X^q - 1 \in \mathbb{F}_p$, A^2 can be easily computed by orthogonality of characters to be

$$A = \begin{pmatrix} \bar{q} & 0 & \cdots & 0 \\ 0 & 0 & \cdots & \bar{q} \\ & & \cdots & \\ 0 & \bar{q} & \cdots & 0 \end{pmatrix}. \tag{1.59}$$

Hence it follows that

$$\delta^2 = |A^2| = (-1)^{\frac{q-1}{2}} \bar{q}^q = \bar{q}^{q-1} \bar{q}^*, \tag{1.60}$$

say, where $q^* = (-1)^{\frac{q-1}{2}} p$. $\frac{q-1}{2}$ being an integer, $\bar{q}^{q-1}$ is a square. Hence $\bar{q}^*$ is a square in $\mathbb{F}_p$ if and only if $\delta \in \mathbb{F}_p$ which is equivalent to $\delta^p = \delta$ by Theorem 1.28, (i).

Now by Theorem 1.28, (i), the raising-to-the-power p map σ is an automorphism. Denoting the rows of A by $\bar{0}, \cdots, \overline{p-1}$ to indicate the exponents of ζ, we find that applying σ to the entries of A has the same effect as multiplying the elements of $\mathbb{F}_q$ by $\bar{p}$. Recalling

$$\mathbb{F}_q^\times / <\bar{p}> = \{\bar{0}, \cdots, \overline{n-1}\},$$

we see that this is the same as the permutation which is the product of $\frac{q-1}{n}$ cycles of length n whose sign is therefore $(-1)^{\frac{q-1}{n}}$. Hence it follows that $\delta^p = (-1)^{\frac{q-1}{n}} \delta$. Hence $\bar{q}^*$ is a square in $\mathbb{F}_p$ if and only if $\frac{q-1}{n}$ is even which is equivalent to the divisibility of $\frac{q-1}{2}$ by n. Then by the Euler criterion, this is equivalent to $\left(\frac{p}{q}\right) = 1$. It follows that $\bar{q}^*$ is a square in $\mathbb{F}_p$, i.e. $\left(\frac{q^*}{p}\right) = 1$ if and only if $\left(\frac{p}{q}\right) = 1$, which amounts to

$$\left(\frac{q^*}{p}\right)\left(\frac{p}{q}\right) = 1. \tag{1.61}$$

Substituting the definition of q^* and appealing to the first supplementary law, we conclude (5.61). $\qquad \square$

Chapter 2

Rudiments of algebraic number theory

In this chapter we give rudiments of algebraic number theory from an elementary standpoint. What we mean is that we present basic results without proof and we illustrate them by concrete examples in the same spirit as in [Kan1]. We assembled some most general results including f.g. modules over a Dedekind domain, which is meant for the integer ring over one in the underlying algebraic number field. In most of the books, mention is made only to the case of f.g. modules over a PID. As a preparation for local theory, we give the completion of algebraic number fields. We state some facts about cyclotomic fields since they are a good example of the theory and could work for an introduction to class field theory.

But the presentation is too sketchy for this chapter to be called an introduction to the theory. The interested reader therefore can consult many existing books on the theory. There are those that can be read more light-heartedly than very serious introductions such as [Lang], [Weil] are [Gold], [Kita], [Nark] etc. with [BShh] as an intermediate. There are many more good books on algebraic number theory and the list is non-exhaustive.

2.1 Galois extensions

Continuing the theory of field extensions in Chapter 1, we shall state rudiments of Galois theory, which describes the structure of a polynomial through the group of permutations of its roots (acting on its splitting field).

Definition 2.1. If G is a subgroup of the group $\text{Aut}\, L$ of all automorphisms (ring isomorphisms) of a field L, then

$$L^G = \{a \in L | a^\sigma = a, \quad \forall \sigma \in G\} \tag{2.1}$$

forms a subfield of L, called the **fixed field** of G. An algebraic extension L/K is called a **Galois extension** if for a subgroup G of $\text{Gal}(L/K)$ of all K-automorphisms of L (ring isomorphisms form L onto L whose restriction to K is the identity),

$$K = L^G.$$

Or we may also say that L/K is Galois if

$$L^\sigma \subset L \text{ for } \forall \sigma \in \mathrm{Gal}(L/K); \quad L^{\mathrm{Gal}(L/K)} = L.$$

This notion is often used in the following context. If an element θ of a Galois extension L of K is fixed by $\forall \sigma \in \mathrm{Gal}(L/K) : \theta^\sigma = \theta$, then $\theta \in K$. For example, if $f \in L[X]$ satisfies $f^\sigma = f$ for $\forall \sigma \in \mathrm{Gal}(L/K)$, then $f \in K[X]$.

Theorem 2.1. *An algebraic extension is a Galois extension if and only if it is a normal, separable extension.*

The following fundamental theorem in Galois theory reduces the study of intermediate fields of a Galois extension to that of subgroups of the Galois group. In particular, it is often used in concluding that if $M = L^G$, then $M = K$.

Theorem 2.2. (Fundamental theorem in Galois theory) *Let L/K be a finite Galois extension with Galois group G. Then the extension degree = the order of the Galois group,*

$$[L : K] = |G|, \tag{2.2}$$

and intermediate fields M of L/K and subgroups H of G are one-to-one correspondence under

$$H = \mathrm{Gal}(L/M), \quad M = L^H, \tag{2.3}$$

and $[L : M] = |H|, [M : K] = (G : H)$.

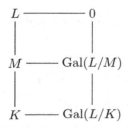

By Theorem 1.21, we have

Corollary 2.1. *If L/K is a separable extension, then there exists a Galois extension L'/K containing L. If $[L : K] < \infty$, then $[L' : K]$ can be taken finite.*

Let L/K be a separable extension of degree n. Take a Galois extension L'/K containing L. By Theorem 1.19, there are n distinct K-isomorphisms $\{\sigma_1, \cdots, \sigma_n\}$. For $\alpha \in L$, define its **characteristic polynomial** F_α by

$$F_\alpha(X) = (X - \alpha^{\sigma_1}) \cdots (X - \alpha^{\sigma_n}). \tag{2.4}$$

Exercise 34. Prove that $F_\alpha \in K[X]$.

Solution. For any $\tau \in \text{Gal}(L/K)$, $\{\tau\sigma_1, \cdots, \tau\sigma_n\} = \{\sigma_1, \cdots, \sigma_n\}$. Hence $F_\alpha^\tau = F$ and so by Theorem 2.2, $F_\alpha \in K[X]$.

Hence, in particular, all symmetric expressions in $\alpha^{\sigma_1}, \cdots, \alpha^{\sigma_n}$ are the elements of K.

Exercise 35. Let $f_\alpha(X) = \text{Irr}(\alpha, K, X)$ be the minimal polynomial of α. Prove that $F_\alpha(X) = f_\alpha(X)^d$, where $d = [L : K(\alpha)]$.

Solution. Since all the roots of $F_\alpha(X)$ are conjugates of α (cf. Corollary 1.6), it follows that all irreducible factors of $F_\alpha(X)$ in $K[X]$ are $f_\alpha(X)$. Hence F_α must be a power of f_α. Since $[K(\alpha) : K] = \deg \text{Irr}(\alpha, K, X)$ and $\deg F_\alpha = [L : K] = n$, it follows that $d[K(\alpha) : K] = n = [L : K]$, whence $d = [L : K(\alpha)]$.

Definition 2.2.

$$\text{Tr}_{L/K}\alpha = \alpha^{\sigma_1} + \cdots + \alpha^{\sigma_n} \in K \tag{2.5}$$

is called the **trace** of α (from L to K) while

$$\text{N}_{L/K}\alpha = \alpha^{\sigma_1} \cdots \cdots \alpha^{\sigma_n} \in K \tag{2.6}$$

is called **norm** of α (from L to K). They appear as coefficients of F_α.

Example 2.1. Let L, K, L' be as above and $[L : K] = n$. Then if $\alpha \in K$, then $\text{Tr}_{L/K}\alpha = n\alpha$ and $\text{N}_{L/K}\alpha = \alpha^n$.

2.2 Modules over Dedekind domains

Definition 2.3. Let M be an Abelian group and let R be a ring with 1. We say that M is a left R-**module** if R acts on M (cf. Definition 1.19) and the action satisfies additivity: if for any $a \in R, x \in M$, we have $ax \in M$ and

(i) $a, b \in R$, $x \in M$, $(ab)(x) = a(bx)$,

(ii) $1x = x$,

(iii) $a \in R$, $x, y \in M$, $a(x + y) = ax + ay$,

(iv) $a, b \in R$, $x \in M$, $(a + b)x = ax + bx$.

If the action is given from the right and it satisfies the counterpart conditions, M is called a right R-module. In what follows we do not distinguish the left and right, and refer to a left R-module as an R-module. Modules over a ring is a generalization of the notion of a vector space over a field and the action corresponds to scalar multiplication. If M is a set and R is a group and conditions (i)-(iii) are satisfied, then R is called a group with an **operator domain** M (or an M-group). Those elements $x \in M$ are called **torsion** elements if for some non-zero $r \in R$, $rx = 0$. The set of all torsion elements form a subgroup of M, called the **torsion subgroup**. An R-module M is said to have a **basis** $\{e_i\} \subset M$ if every element $x \in M$ can be expressed unique as $x = \sum_i r_i e_i$, where r_i are 0 except for finitely many i. If M has a basis, it is called a **free** R-module. A free $\mathbb{Z}$-module is called a **free Abelian group**.

Let R be an integral domain and K its quotient field. An R-submodule $\mathfrak{a}$ of K is called a **fractional ideal** if there is a non-zero element $r \in R$ so that $r\mathfrak{a} \subset R$ (if $r \in K$, we may clear the denominator by multiplying an element of R). In contrast to fractional ideals, ordinary ideals of R are called **integral ideals**.

Proposition 2.1. *Two fractional ideals* $\mathfrak{a}, \mathfrak{b}$ *are isomorphic as R-modules if and only there exists an $r \in K$ such that* $\mathfrak{a} = r\mathfrak{b}$.

Proof.   Plainly $\mathfrak{a} \simeq r\mathfrak{a}$. It suffices therefore to prove the converse for integral ideals. For a homomorphism $f : \mathfrak{a} \to \mathfrak{b}$, $a^{-1}f(a)$ $(0 \neq a \in \mathfrak{a})$ is constant. Indeed, for another $0 \neq a' \in \mathfrak{a}$, $af(a') = f(a'a) = f(aa') = a f(a')$. Hence putting $a^{-1}f(a) = r$, we have $f(a) = ra$, which gives an isomorphism, completing the proof. $\square$

For a fractional ideal $\mathfrak{a}$, let $\mathfrak{a}^{-1} = \{x \in K | x\mathfrak{a} \subset R\}$ be the **inverse fractional ideal** of $\mathfrak{a}$. $\mathfrak{a}^{-1}\mathfrak{a} \subset R$ holds.

Exercise 36. Prove that $\mathfrak{a}^{-1}$ is a fractional ideal.

Solution.   We may easily check that $\mathfrak{a}'$ is an Abelian group. It therefore suffices to find an element to clear the denominators. Choose a non-zero element $r \in R$ such that $r\mathfrak{a} \subset R$. Then for any $0 \neq a \in \mathfrak{a}$, $ra \in R$ and $ra \in \mathfrak{a}$ since $\mathfrak{a}$ is an R-module. Hence $ra \in R \cap \mathfrak{a}$. Hence for any $x \in \mathfrak{a}^{-1}$, $rax \in R$ by definition. Hence $ra\mathfrak{a}^{-1} \subset R$.

Definition 2.4. If $\mathfrak{a}^{-1}\mathfrak{a} = R$ holds, $\mathfrak{a}$ is called **invertible**. A domain is called a **Dedekind domain** if all its fractional ideals are invertible. An element of an extension domain S of a domain R is called **integral** over R if it is a root of a monic polynomial $\in R[X]$. If all integral elements in the quotient field K of R are only those in R, then R is said to be **integrally closed**. A ring is called **Noetherian** if one of the following equivalent conditions holds true.

(i) (maximal condition) In any set of ideals in R there exists a maximal element.

(ii) (ascending chain condition) Any infinite chain of ideals

$$\mathfrak{a}_1 \subset \mathfrak{a}_2 \subset \cdots$$

terminates, i.e. from some number N on, they are all equal.

(iii) Any ideal of R are finitely generated.

Theorem 2.3. *A Dedekind domain is characterized by one of the following two equivalent conditions.*
(i) It is an integrally closed, Noetherian domain in which every prime ideal is maximal.
(ii) Any fractional ideal may be expressed uniquely as a product of prime ideals $\mathfrak{p}_1^{e_1} \cdots \mathfrak{p}_1^{e_\nu}$, $e_i \in \mathbb{Z}$.

In view of Proposition 2.1, it is natural to classify modules according to the isomorphism. Each equivalence class of isomorphic ideals is called an **ideal class**. They form a group called the **ideal class group**. Its order is called the **class number**.

Exercise 37. Prove the following assertions. (i) In a Dedekind domain, the set I of all fractional ideals form a group and the set P of all principal ideals forms its subgroup.
(ii) The ideal class group is isomorphic to the factor group I/P. A PID is a Dedekind domain whose ideal class group is the unit group.

Example 2.2. Consider the imaginary quadratic field $\mathbb{Q}(\sqrt{-5}) = \{a + b\sqrt{-5} | a, b \in \mathbb{Q}\}$. Its ring of integers is given by $\mathbb{Z}(\sqrt{-5}) = \{a + b\sqrt{-5} | a, b \in \mathbb{Z}\}$. We illustrate by the decomposition $6 = 2 \cdot 3 = (1 - \sqrt{-5})(1 + \sqrt{-5})$ and show that $\mathbb{Z}(\sqrt{-5})$ is not a PID. Indeed, since $N2 = 4$, $N3 = 9$, $N(1 \pm \sqrt{-5}) = 6$, these cannot be associates. Thus we are to consider a substitute for ordinary prime decomposition. And it is the decomposition into prime ideals given by Theorem 2.3.

In the times of E. Kummer, it was thought that all rings of integers are UFD and Kummer thought that he proved Fermat's last theorem. Kummer's investigation depended on the UF property of rings and does not apply to all rings. To avoid this difficulty, he introduced the idea of **ideal numbers**, which eventually got developed into modern theory of ideals by Dedekind (L. Kronecker, E. Noether). Algebraic number theory has been developed in order to assure the *prime decomposition.*

We quote the following fundamental theorem on a finitely generated module over a Dedekind domain (cf. [Nark, Theorems 1.10 and 1.11]).

Theorem 2.4. *Let R be a Dedekind domain and let M be a finitely generated R-module. Then M is isomorphic to*

$$R^k \oplus \mathfrak{i} \oplus A,$$

where A is the torsion subgroup of M, $k \in \mathbb{N}$ and $\mathfrak{i}$ is some fractional ideal of R. The expression is unique up to orders of entries.

In most of books, the above theorem is stated for the case of R being a *PID*. Since we consider only algebraic number fields over $\mathbb{Q}$, the ring of integers is a free $\mathbb{Z}$-module, and (2.7) applies (cf. also Chapter 4, Lemma 4.3). When we consider relative extensions, we are to appeal to Theorem 2.4. Indeed, what is more interesting is the case of relative extensions. Indeed, the celebrated class field theory is a theory of relative Abelian extensions. For Abelian extensions over $\mathbb{Q}$ we may apply the Kronecker-Weber Theorem (Theorem 2.14) and we may make detailed analysis. Similarly, class field theory enables us to make a detailed study on relative Abelian extensions.

Theorem 2.5. *Let R be a PID and let M be a finitely generated R-module. Then M is isomorphic to*

$$R/e_1R \oplus \cdots \oplus R/e_lR$$

where e_i may be chosen so that $e_i|e_{i+1}$ $(1 \leq i \leq l-1)$ and in choosing so, they are uniquely determined up to associates. Or more concretely,

$$\mathbb{Z}/a_1\mathbb{Z} \oplus \cdots \oplus \mathbb{Z}/a_l\mathbb{Z} \oplus \underbrace{\mathbb{Z} \oplus \cdots \oplus \mathbb{Z}}_{r \text{ times}}, \tag{2.7}$$

where $1 < a_1$, $a_i|a_{i+1}$ $(1 \leq i \leq l-1)$ and the product $a = a_1 \cdots a_l$ and r are uniquely determined.

$e_1, \cdots, e_l$ or $a_1, \cdots, a_l$ are called **elementary divisors**. r is called the R-**rank** of G.

If k is a finite extension over $\mathbb{Q}$ contained in its algebraic closure $\bar{\mathbb{Q}}$, then it is called an **algebraic number field** or simply a **number field**. By Corollary 1.5, it is a finitely generated algebraic extension. All the integral elements over $\mathbb{Q}$ in k forms a Dedekind domain $\mathcal{O}_k$ (also often denoted by $\mathfrak{o}$) and is called the **ring of algebraic integers**. The notation originates from the term "order" ([BShh, Definition, p.88]) (eine "Ordnung" in German). Then k is the quotient field of $\mathcal{O}_k$. $\mathcal{O}_k$ is a Dedekind ring and so the prime decomposition holds true in the sense of Theorem 2.3, (ii), i.e. every integral ideal $\mathfrak{a}$ may be expressed uniquely as a product of prime ideals $\mathfrak{a} = \mathfrak{p}_1^{e_1} \cdots \mathfrak{p}_1^{e_\nu}$, $0 \le e_i \in \mathbb{Z}$. Therefore we may speak of the gcd, lcm just as in the ring of rational integers, which is UFD, being Euclidean.

By Remark 2.1, every ideal has a finite norm. Thus algebraic number theory may be described in one aspect as *a study on modules over a Dedekind domain with finite norm property.*

Exercise 38. (i) If $\mathfrak{a} = \prod_{i=1}^m \mathfrak{p}_i^{a_i}$ and $\mathfrak{b} = \prod_{i=1}^m \mathfrak{p}_i^{b_i}$, $0 \le a_i, b_i \in \mathbb{Z}$, then

$$\mathfrak{a} + \mathfrak{b} = \prod_{i=1}^m \mathfrak{p}_i^{\min\{a_i, b_i\}}, \quad \mathfrak{a} \cap \mathfrak{b} = \prod_{i=1}^m \mathfrak{p}_i^{\max\{a_i, b_i\}}$$

(cf. Remark 1.2).

(ii) Every fractional ideal $\mathfrak{a}$ may be given in the form $\mathfrak{a} = \prod_{i=1}^m \mathfrak{p}_i^{a_i} \prod_{i=1}^n \mathfrak{p}_i^{b_i}$, $0 \le a_i, b_i \in \mathbb{Z}$. Prove that

$$\mathfrak{a} \cap \mathcal{O}_k = \mathfrak{a} \cap \mathfrak{b}. \tag{2.8}$$

2.3 Algebraic number fields

Let k be a number field. Since char $\mathbb{Q} = 0$, it follows from Theorem 1.20 that k is a simple extension $k = \mathbb{Q}(\theta)$. Hence we may apply Theorem 1.15: If $f(X) = \mathrm{Irr}(\theta, X, \mathbb{Q})$ with $\deg f = n$, then $k = \mathbb{Q} \oplus \mathbb{Q}\theta \oplus \cdots \oplus \mathbb{Q}\theta^{n-1}$.

We may suppose θ has r_1 real conjugate roots $\theta^{(j)}$, $1 \le j \le r_1$ and $2r_2$ imaginary conjugate roots $\theta^{(j)}, \overline{\theta^{(j)}}$, $1 \le i \le r_2$. Here $n = r_1 + 2r_2$. Then there arise corresponding r_1 real conjugates $k^{(j)}$, $1 \le j \le r_1$ and $2r_2$ imaginary conjugates $k^{(j)}, \overline{k^{(j)}}$, $1 \le i \le r_2$. If they all coincide, then $k/\mathbb{Q}$ is a normal extension and so by Theorem 2.1, it is a Galois extension.

Lemma 2.1. *Let k be an algebraic number field. Define a bilinear form $B_k(x, y)$ on $k \times k$ by $\mathrm{Tr}(xy)$ for $x, y \in k$, where $\mathrm{Tr} = Tr_{F/\mathbb{Q}}$. If $B_k(x, y) = 0$ for all $y \in k$, then we have $x = 0$, and if $\omega_1, \cdots, \omega_n$ is a basis of k over $\mathbb{Q}$, then the matrix $(B_F(\omega_i, \omega_j))$ is regular.*

Proof. Suppose that $x \in k$ is not the zero and $B_F(x, y) = 0$ for all $y \in F$. Putting $y = 1/x$, we have the contradiction $[F : \mathbb{Q}] = Tr(1) = Tr(x \cdot 1/x) = 0$. If the rational matrix $(B_k(\omega_i, \omega_j))$ is not regular, there is a non-zero rational vector $\mathbf{x} = (x_1, \cdots, x_n) \in \mathbb{Q}^n$ such that $\mathbf{x}(B_F(\omega_i, \omega_j)) = (B_F(\sum x_i \omega_i, \omega_1), \cdots, B_F(\sum x_i \omega_i, \omega_n)) = (0, \cdots, 0)$, whence $B_F(\sum x_i \omega_i, y) = 0$ for all $y \in F$. This and the former part imply the contradiction $\sum x_i \omega_i = 0$, i.e. $x_1 = \cdots = x_n = 0$. $\square$

Lemma 2.2. *Let $\omega_1, \cdots, \omega_n$ be a basis of F over $\mathbb{Q}$. Then there is a basis $\omega_1', \cdots, \omega_n'$ of F over $\mathbb{Q}$ such that*

$$B_F(\omega_i, \omega_j') = \delta_{ij} = \begin{cases} 1 & \text{if } i = j, \\ 0 & \text{otherwise.} \end{cases}$$

Proof. Put $M = (B_F(\omega_i, \omega_j))$ and $(\omega_1', \cdots, \omega_n') = (\omega_1, \cdots, \omega_n)M^{-1}$. Writing $M^{-1} = (m_{ij})$, we have

$$B_F(\omega_i, \omega_j') = B_F(\omega_i, \sum_k \omega_k m_{kj}) = \sum_k B_F(\omega_i, \omega_k)m_{kj} = \delta_{ij}.$$

If $\sum_j a_j \omega_j' = 0$ for $a_j \in \mathbb{Q}$, then we have $a_i = B_F(\omega_i, \sum_j a_j \omega_j') = 0$. Thus $\omega_1', \cdots, \omega_n'$ are linearly independent over $\mathbb{Q}$. $\square$

$\omega_1', \cdots, \omega_n'$ is called the **dual basis** of $\omega_1, \cdots, \omega_n$.

Lemma 2.3. *If $\alpha \in \mathcal{O}_k$. Then we have $N_{F/\mathbb{Q}}\, \alpha \in \mathbb{Z}$ and $Tr_{F/\mathbb{Q}}\, \alpha \in \mathbb{Z}$.*

Proof. By definition, the minimal polynomial of $\alpha \in \mathcal{O}$ is in $\mathbb{Z}[X]$. Since these numbers appear as its coefficients (cf. Definition 2.2), the assertion follows. $\square$

Recall that a $\mathbb{Z}$-module is called a free $\mathbb{Z}$-module of rank n if it is isomorphic to $\mathbb{Z}^n$.

Theorem 2.6. *Let N be a free $\mathbb{Z}$-module of rank n. Then a subgroup M of N is a free $\mathbb{Z}$-module of rank $m\,(\leq n)$.*

Proof. Let $v_1, \cdots, v_n$ be a basis of N and put

$$N_k = \{a_1 v_1 + \cdots + a_k v_k \mid a_1, \cdots, a_k \in \mathbb{Z}\} \text{ and } M_k = M \cap N_k;$$
$$M_n = M, \ M_0 = \{0\}.$$

Since

$$I_k = \{a_k \in \mathbb{Z} \mid a_1 v_1 + \cdots + a_k v_k \in M\}$$

is a submodule of $\mathbb{Z}$, there is an integer $b_k\,(\geq 0)$ such that $I_k = b_k\mathbb{Z}$ (by Theorem 1.5). Take an element $u_k = b_{1,k}v_1 + \cdots + b_{k,k}v_k \in M$ $(b_{k,k} = b_k)$. Clearly $u_k \in M_k$. If $b_k = 0$, then we choose 0 as u_k. Then for any element $v \in M_k$, there is an integer c such that

$$v - cu_k \in M_{k-1}\,(= M \cap N_{k-1}).$$

Let $v \in M$. Applying the above argument to v in the descending order, we see that there are integers $c_n, \cdots, c_1$ such that

$$v - c_nu_n - \cdots - c_ku_k \in M_{k-1}\ (k = n, n-1, \cdots, 1).$$

Thus we have $M = \{c_1u_1 + \cdots + c_nu_n \mid c_1, \cdots, c_n \in \mathbb{Z}\}$.

Let $u_{i_1}, \cdots, u_{i_m}$ $(1 \leq i_1 < \cdots < i_m \leq n)$ be all non-zero elements in $\{u_1, \cdots, u_n\}$. Then we have

$$M = \{c_{i_1}u_{i_1} + \cdots + c_{i_m}u_{i_m} \mid c_{i_1}, \cdots, c_{i_m} \in \mathbb{Z}\}.$$

Next, suppose that $c_{i_1}u_{i_1} + \cdots + c_{i_\ell}u_{i_\ell} = 0$ with $c_{i_\ell} \neq 0$; then the definition of N_k implies

$$c_{i_\ell}u_{i_\ell} \notin N_{i_\ell-1}.$$

On the other hand, we have

$$c_{i_\ell}u_{i_\ell} = -(c_{i_1}u_{i_1} + \cdots + c_{i_{\ell-1}}u_{i_{\ell-1}}) \in N_{i_{\ell-1}} \subset N_{i_\ell-1},$$

by $i_{\ell-1} \leq i_\ell - 1$. This is a contradiction, and so $u_{i_1}, \cdots, u_{i_m}$ $(1 \leq i_1 < \cdots < i_m \leq n)$ is a basis of M. $\qquad \square$

Remark 2.1. A slight modification of Theorem 2.6 shows that the group index $(\mathcal{O} : \mathfrak{a})$ is finite $(\in \mathbb{N} \cup \{0\})$ for any integral ideal $\mathfrak{a} \subset \mathcal{O}$. It is denoted by $N\mathfrak{a}$ and called the **norm** of $\mathfrak{a}$:

$$N\mathfrak{a} = (\mathcal{O} : \mathfrak{a}). \tag{2.9}$$

Theorem 2.7. *Let k be an algebraic number field. Then $\mathcal{O}_k$ is a free $\mathbb{Z}$-module of rank $n = [k : \mathbb{Q}]$, i.e. there are elements $\omega_1, \cdots, \omega_n \in k$ such that*

$$\mathcal{O}_k = \mathbb{Z}\omega_1 \oplus \cdots \oplus \mathbb{Z}\omega_n.$$

Proof. Take an element $\theta \in F$ such that $F = \mathbb{Q}(\theta)$. Moreover we may assume $\theta \in \mathcal{O}_k$ by clearing the denominator. Since $u_i = \theta^{i-1}$ $(i = 1, \cdots, n)$ is a basis of F over $\mathbb{Q}$, we can take a dual basis $u_1', \cdots, u_n'$ by Lemma 2.2. Let $w \in \mathcal{O}_k$ and $w = \sum a_iu_i'$ $(a_i \in \mathbb{Q})$. Since $w, u_i \in \mathcal{O}_k$, we have $a_i = B_F(w, u_i) \in \mathbb{Z}$ by Lemma 2.3. Thus $\mathcal{O}_k$ is a submodule of $\mathbb{Z}u_1' + \cdots + \mathbb{Z}u_n'$. Theorem 2.6 implies that $\mathcal{O}_k$ has a basis $\omega_1, \cdots, \omega_m$ $(m \leq n)$.

Since $\mathcal{O}_k \subset \mathbb{Z}\omega_1 + \cdots + \mathbb{Z}\omega_m$, we have $F \subset \mathbb{Q}\omega_1 + \cdots + \mathbb{Q}\omega_m$, i.e. $n = \dim_{\mathbb{Q}} F \leq m$. Thus we obtain $n = m$. $\qquad \square$

We call $\omega_1, \cdots, \omega_n$ a $\mathbb{Z}$-basis of $\mathcal{O}_k$ and often write

$$\mathcal{O}_k = \mathbb{Z}[\omega_1, \cdots, \omega_n].$$

Definition 2.5. A finitely generated $\mathbb{Z}$-module M in a number filed k of degree n is called a **full module** if it contains n linearly independent elements over $\mathbb{Q}$. This amounts to the existence of $\mathbb{Q}$-basis $\alpha_1, \cdots, \alpha_n$ such that

$$M = \mathbb{Z}\alpha_1 \oplus \cdots \oplus \mathbb{Z}\alpha_n,$$

which is called an integral basis relative to M.

The Chinese remainder theorem (Theorem 1.23 in Chapter 1) may be generalized as

Lemma 2.4. *Let* $\mathfrak{a}_1, \cdots, \mathfrak{a}_m$ *be mutually relatively prime integral ideals. Then there are elements* $a_i \in \mathcal{l}_k$ $(1 \leq i \leq m)$ *such that*

$$a_i \equiv \begin{cases} 1 \bmod \mathfrak{a}_i, \\ 0 \bmod \mathfrak{a}_j & \text{if } j \neq i. \end{cases} \tag{2.10}$$

Proof. If $j \neq i$, then $\mathfrak{a}_i + \mathfrak{a}_j = \mathcal{O}_k$ implies that there are $b_i \in \mathfrak{a}_i$ and $b_j \in \mathfrak{a}_j$ such that $b_i + b_j = 1$. Therefore we have

$$b_j \equiv \begin{cases} 1 \bmod \mathfrak{a}_i, \\ 0 \bmod \mathfrak{a}_j. \end{cases}$$

Thus $a_i = \prod_{j \neq i} b_j$ satisfies the assertion of the lemma.   □

Theorem 2.8. *Let* $\mathfrak{a}_1, \cdots, \mathfrak{a}_m$ *be mutually relatively prime integral ideals, and let* $x_1, \cdots, x_m \in \mathcal{O}_k$. *Then there is an element* $x \in \mathcal{O}_k$ *such that*

$$x \equiv x_i \bmod \mathfrak{a}_i \quad (i = 1, \cdots, m).$$

Proof. Let a_i be that in the previous lemma. Then $x = \sum x_i a_i$ satisfies the assertion.   □

This theorem is called the (weak) approximation theorem. With this, Proposition 1.9 can be generalized in the following form.

Proposition 2.2. *Let* $\mathfrak{a}$ *and* $\mathfrak{b}$ *be integral ideals of* F. *Then we have*

$$\mathcal{O}_k/\mathfrak{b} \simeq \mathfrak{a}/\mathfrak{a}\mathfrak{b},$$

in particular

$$(\mathcal{O}_k : \mathfrak{b}) = (\mathfrak{a} : \mathfrak{a}\mathfrak{b}).$$

Proof. Let $\mathfrak{a} = \prod_{i=1}^{m} \mathfrak{p}_i^{a_i}$ and $\mathfrak{b} = \prod_{i=1}^{m} \mathfrak{p}_i^{b_i}$ $(a_i, b_i \geq 0)$, where $\mathfrak{p}_i$ are distinct prime ideals. We take $\alpha \in \mathcal{O}_k$ such that

$$\alpha \in \mathfrak{p}_i^{a_i} \setminus \mathfrak{p}_i^{a_i+1} \text{ for } 1 \leq i \leq m, \tag{2.11}$$

using Theorem 2.8. Then we shall show that

$$\mathfrak{a}\mathfrak{b} + (\alpha) = \mathfrak{a}. \tag{2.12}$$

Since $\mathfrak{a}\mathfrak{b} + (\alpha) \subset \mathfrak{p}_i^{a_i}$ for all i, we have $\mathfrak{a}\mathfrak{b} + (\alpha) \subset \mathfrak{a}$. If $\mathfrak{a}\mathfrak{b} + (\alpha) \neq \mathfrak{a}$, then there is a prime ideal $\mathfrak{p}$ such that $\mathfrak{a}\mathfrak{b} + (\alpha) \subset \mathfrak{a}\mathfrak{p}$. We have $\prod \mathfrak{p}_i^{a_i+b_i} = \mathfrak{a}\mathfrak{b} \subset \mathfrak{a}\mathfrak{b} + (\alpha) \subset \mathfrak{a}\mathfrak{p} \subset \mathfrak{p}$, which yields $\mathfrak{p} \supset \mathfrak{p}_i$ for some i. Hence $\mathfrak{p} = \mathfrak{p}_i$ follows and then $\alpha \in \mathfrak{a}\mathfrak{p}_i \subset \mathfrak{p}_i^{a_i+1}$ holds, which contradicts (2.11), thereby showing (2.12).

Consider the translation

$$f : \mathcal{O}_k \to \mathfrak{a}/\mathfrak{a}\mathfrak{b}; \quad f(a) = a\alpha \tag{2.13}$$

which is well-defined and surjective by (2.12).

Note that $\operatorname{Ker} f$ consists of those a's, for which $a\alpha \in \mathfrak{a}\mathfrak{b}$, or $a \in \alpha^{-1}\mathfrak{a}\mathfrak{b}$. Since (2.11) implies $(\alpha) = \mathfrak{a} \prod \mathfrak{q}_i^{c_i}$, where prime ideals $\mathfrak{q}_i$ are different from any $\mathfrak{p}_j$ and $c_i \geq 0$, it follows that $a \in (\mathfrak{a} \prod \mathfrak{q}_i^{c_i})^{-1}\mathfrak{a}\mathfrak{b}$, and so $a \in \mathfrak{b} \prod \mathfrak{q}_i^{-c_i} \cap \mathcal{O}_k = \mathfrak{b}$ by (2.8). $\operatorname{Ker} f \supset \mathfrak{b}$ being clear, we have $\operatorname{Ker} f = \mathfrak{b}$ and so Theorem 1.10 applies. $\qquad \square$

Remark 2.2. Proposition 2.2 is similar to Theorem 1.11. Application of Proposition 2.2 yields multiplicativity of norms (cf. Remark 1.4). The Dirichlet series

$$\zeta_k(s) = \sum_{\mathfrak{a} \neq \mathfrak{o}} \frac{1}{N\mathfrak{a}^s} = \sum_{n=1}^{\infty} \frac{F(n)}{n^s}, \quad \sigma > 1 \tag{2.14}$$

is called the **Dedekind zeta-function** and $F(n) = \sum_{N\mathfrak{a}=n} 1$ is the number of ideals with norm n, called the Idealfunktion (ideal function). The Dedekind zeta-functions play equally important role in algebraic number theory as does the Riemann zeta-function in rational number field.

We introduce the notion of the (absolute) **discriminant**. Let $w_1, \cdots, w_n$ be a $\mathbb{Z}$-basis of $\mathcal{O}_k$. The (absolute) **discriminant**, denoted by D_k, is defined by

$$D_k = \det(Tr_{K/\mathbb{Q}}(w_i w_j)),$$

where $Tr = Tr_{K/\mathbb{Q}}$. It is a non-zero integer (cf. Lemmas 2.1 and 2.3). It is independent of the choice of a $\mathbb{Z}$-basis. For let $w_1', \cdots, w_n'$ be another $\mathbb{Z}$-basis of $\mathcal{O}_k$; then there is a matrix $U = (u_{ij})$ $(u_{ij} \in \mathbb{Z})$ such that

$$(w_1', \cdots, w_n') = (w_1, \cdots, w_n)U, \ \det U = \pm 1.$$

It is easy to see that
$$(Tr_{K/\mathbb{Q}}(w'_i w'_j)) = {}^t U (Tr_{K/\mathbb{Q}}(w_i w_j)) U,$$
whence $\det(Tr_{K/\mathbb{Q}}(w'_i w'_j)) = \det(Tr_{K/\mathbb{Q}}(w_i w_j))$.

2.4   Completion of a number field

Let k be an algebraic number field.

Definition 2.6. If a non-negative-valued function $v : k \to \mathbb{R}_+ \cup \{0\}$ satisfies the following conditions, it is called a **valuation** or an **absolute value** on k and denoted $|\cdot|$:

(i) $v(0) = 0$; $v(x) = 0 \Longleftrightarrow x = 0$.

(ii) $v(xy) = v(x)v(y)$.

(iii) There exists a $C > 0$ such that $v(x + y) \leq C \max\{v(x), v(y)\}$.

If (iii) holds with $C = 1$, then v is said to be **non-Archimedean** (null-A) or finite. Otherwise, it is said to be **Archimedean** or infinite.

Exercise 39. Prove that $C = 2$ in (ii) is equivalent to the **trigonometric inequality**
$$v(x + y) \leq v(x) + v(y). \tag{2.15}$$

Hereafter, we assume (39) and write $|x + y| \leq |x| + |y|$. Needless to say, it induces a metric $d(x, y) = |x - y|$ on k. We sometimes use both notation. We also exclude the trivial valuation $v(0) = 0, v(x) = 1$, $x \neq 0$. There exist infinitely many non-Archimedean valuations corresponding to infinitely many prime ideals in k and $[k : \mathbb{Q}]$ Archimedean valuations corresponding to $[k : \mathbb{Q}]$ embeddings of k into $\mathbb{C}$. Non-Archimedean valuations define the "nearness" by the degree of divisibility of two elements by a power of the prime ideal.

Example 2.3. For an $x \in \mathbb{Q}$ and a prime p, we define the **p-adic valuation** $v_p(x)$ of x (also called p-order) by
$$x = p^{v_p(x)} \frac{a}{b}, \quad a, b \in \mathbb{Z}, \ (a, p) = (b, p) = 1. \tag{2.16}$$
Then the p-adic metric φ_p is defined by $\varphi_p(x) = \rho^{v_p(x)}$ for some $0 < \rho < 1$. We usually make the normalization $\rho = \frac{1}{p}$ so that the product formula holds true. Cf. [BShh, p. 34].

Definition 2.7. $\tilde{k}$ is called a **completion** of k with respect to the valuation v if

(i) $\tilde{k}$ is complete.

(ii) There exists an isometry $\rho : k \to \tilde{k}$ such that $\overline{\rho(k)} = \tilde{k}$.

Our objective is to prove the following theorem, thereby constructing the real numbers simultaneously.

Theorem 2.9. *For every (non-trivial) valuation v of k there exists a unique (up to isometry) completion k_v of k.*

2.4.1 *Construction of the field*

A sequence $\{a_n\} \in k$ is called a **Cauchy sequence** if the following holds.

$$\forall \varepsilon > 0, \exists n_0 = n_0(\varepsilon) \in \mathbb{N} \text{ such that for } m, n > n_0, |a_n - a_m| < \varepsilon. \quad (2.17)$$

Exercise 40. Let $\mathcal{R} = \{\{a_v\}\}$ be the set of all Cauchy sequences in k. Then prove that $(\mathcal{R}, +, \cdot)$ forms a commutative ring.

Proof. We define addition and multiplication componentwise and show that they satisfy the ordinary defining conditions for a ring. $\{a_n\} + \{b_n\} = \{a_n + b_n\}$, $\{a_n\}\{b_n\} = \{a_n b_n\}$.

(1). We need to show that $\{a_n + b_n\}$ is a Cauchy sequence. For two Cauchy sequences $\{a_n\}, \{b_n\}$ we may choose n_0 in (2.17) in common and assert that

$$\forall \varepsilon > 0, \exists n_0 \in \mathbb{N} \text{ such that for } m, n \geq n_0, |a_n - a_m| < \varepsilon, |b_n - b_m| < \varepsilon. \quad (2.18)$$

Hence $|a_n + b_n - a_m - b_m| \leq |a_n - a_m| + |b_n - b_m| < 2\varepsilon$.

(2). $\{0\} = \{0, 0, \cdots\} \in \mathcal{R}$ is the identity. To show that $\{0\} \in \mathcal{R}$ we note that *a convergent sequence is a Cauchy sequence.* Indeed, if $\lim_{n \to \infty} a_n = a$, then for $\forall \varepsilon > 0$ $\exists n_0 \in \mathbb{N}$ such that for $n \geq n_0$, $d(a_n, a) < \varepsilon$. Hence, for $m, n \geq n_0$, $d(a_n, a_m) \leq d(a_n, a) + d(a, a_m) < 2\varepsilon$.

(3). For each $\{a_n\} \in \mathcal{R}$, $\{-a_n\} \in \mathcal{R}$ is its inverse element.

(4). Clearly, the addition is commutative.

(5). We first show that *a Cauchy sequence is bounded.* In Equation (2.17), choose $\varepsilon = 1$ and $m = n_0$. Then for $n \geq n_0$, $|a_n| \leq |a_{n_0}| + 1$. Putting $M = \max\{|a_1|, \cdots, |a_{n_0 - 1}|, |a_{n_0}| + 1\}$, we obtain $|a_n| \leq M$ for all

$n \in \mathbb{N}$. Hence from (2.17) it follows that $|a_n b_n - a_m b_m| = |a_n b_n - a_n b_m + a_n b_m - a_m b_m| \leq (|a_n| + |b_m|)\varepsilon$.

(6).   Distributive law

and

(7).   Commutative law clearly hold. □

Lemma 2.5. *Let* $\mathcal{I} = \{\{a_n\} | a_n \to 0 \ (n \to \infty)\} \subset \mathcal{R}$ *be the set of sequences converging to 0, which are Cauchy sequences by the remark above. Then* $\mathcal{I}$ *is a maximal ideal of* $\mathcal{R}$.

Proof.   Since the subtraction $\{x_n\} - \{y_n\}\{x_n - y_n\} \in \mathcal{I}$, $\mathcal{I}$ is plainly an additive subgroup of $\mathcal{R}$. Since a Cauchy sequence is bounded, multiplication of any element of $\mathcal{I}$ by an element of $\mathcal{R}$ remains within $\mathcal{I}$, whence $\mathcal{I}$ is an ideal.

Next let $y = \{y_n\} \in \mathcal{R} \backslash \mathcal{I}$. Then we show that y^{-1} exists in $\mathcal{I}$. For this we show that $\{y_n\}$ is bounded from below, i.e. $|y_n| \geq \delta$ for all $n \in \mathbb{N}$. We use (2.17) in the form with a fixed $\varepsilon > 0$, $\exists n_0(\varepsilon) \in \mathbb{N}$ such that $|y_n - y_{n_0}| < \varepsilon$. Choose $0 < \delta < \varepsilon$. Then for $n \geq n_0$, $|y_n| - |y_{n_0}| \leq |y_n - y_{n_0}| < \varepsilon$, whence $|y_n| \geq |y_{n_0}| - \varepsilon > \delta - \varepsilon > 0$. We may add e.g $\{1, 1, \cdots, 1, 0, 0, \cdots\} \in \mathcal{I}$ to $\{y_n\}$ to ensure that all the terms of $\{y_n\}$ are non-zero. Hence we obtain $\{y_n^{-1}\} \in \mathcal{R}$. Then the ideal $< \mathcal{I}, y >$ generated by $\mathcal{I}$ and y contains 1 and must coincide with $\mathcal{R}$. Since any ideal $\mathcal{J}$ containing $\mathcal{I}$ contains $< \mathcal{I}, y >$, it follows that $\mathcal{I} = \mathcal{R}$, whence $\mathcal{I}$ is maximal. □

2.4.2   *Proof of completeness*

Now that we have constructed a field $\tilde{k} = \mathcal{R}/\mathcal{I}$, we introduce the embedding

$$\rho : k \to \tilde{k}; \quad x \to \{x, x, \cdots\} + \mathcal{I}, \tag{2.19}$$

which is a field injection and we may view k as a subfield of $\tilde{k}$. We now assume k to be an ordered field (as $\mathbb{Q}$) and introduce the ordering on $\tilde{k}$. We define $\tilde{x} > \tilde{0}$ if there exists an $k \ni \varepsilon > 0$ and a representative $\{x_n\} \subset k$ $(\tilde{x} = \{x_n\} + \mathcal{I})$ such that $x_n \geq \varepsilon$. We assume that the Archimedean principle holds true in k: For any $a > 0$, $\varepsilon > 0$, there is an $n_0 \in \mathbb{N}$ such that

$$n_0 \varepsilon > a.$$

Then we may introduce the absolute values on $\tilde{k}$ and therewith a distance d on $\tilde{k}$: $\tilde{d}(\tilde{x}, \tilde{y}) = |\tilde{x} - \tilde{y}|$. We may define the limit notion and Cauchy sequences etc. in exactly the same way as in calculus: for example, a sequence $\{\tilde{x}_n\} \subset \tilde{k}$ is a Cauchy sequence if and only if for any $k \ni \varepsilon > 0$, there exists an $n_0 \in \mathbb{N}$ such that for any $\mathbb{N} \ni m, n \geq n_0$, we have

$$\tilde{d}(\tilde{x}_m, \tilde{x}_n) < \varepsilon. \tag{2.20}$$

We note that since $d(x_m, x_n) = \tilde{d}(\rho(x_m), \rho(x_n)) < \varepsilon$ for $m, n \geq n_0$, $\tilde{x} = \{x_n\} + \mathcal{I}$ is equivalent to

$$\tilde{d}(\tilde{x}, \rho(x_n)) < \varepsilon, \tag{2.21}$$

i.e. $x_n = \rho(x_n) \to \tilde{x}$. Hence any element $\tilde{x} \in \tilde{k}$ is the limit of a sequence of k.

Theorem 2.10. $\tilde{k}$ *is a complete metric space and k is its dense subset.*

Proof. Let $\{\tilde{x}_n\} \subset \tilde{k}$ be a Cauchy sequence. Then by (2.21), we conclude that for each $n \in \mathbb{N}$, there is a $y_n \in k$ such that

$$\tilde{d}(\tilde{x}_n, \rho(y_n)) < \frac{1}{n}. \tag{2.22}$$

The $\{y_n\}$ is a Cauchy sequence. For choosing n_0 so large that $n_0 \varepsilon \geq 1$, we have for $m, n \geq n_0$

$$d(y_m, y_n) = \tilde{d}(\rho(y_m), \rho(y_n)) \leq \tilde{d}(\rho(y_m), \tilde{x}_m) + \tilde{d}(\tilde{x}_m, \tilde{x}_n) + \tilde{d}(\tilde{x}_n, \rho(y_n)) \tag{2.23}$$

$$< \frac{1}{n} + \varepsilon + \frac{1}{n} < 3\varepsilon.$$

Hence $\{y_n\} \in \mathcal{R}$ and it defines a class $\tilde{y} = \{y_n\} + \mathcal{I}$. Since $\tilde{d}(\rho(y_n), \tilde{y}) \to 0$ as $n \to \infty$, it follows that

$$\tilde{d}(\tilde{x}_n, \tilde{y}) \leq \tilde{d}(\tilde{x}_n, \rho(y_n)) + \tilde{d}(\rho(y_n), \tilde{y}) < \frac{1}{n} + \varepsilon, \tag{2.24}$$

whence $\tilde{x}_n \to \tilde{y}$, i.e. every Cauchy sequence is convergent. $\qquad\square$

Remark 2.3. In the case $k = \mathbb{Q}$, the completion $\mathbb{R} = \mathcal{R}/\mathcal{I}$ is a maximal Archimedean ordered field.

2.5 The quadratic field with the golden section unit

In this section we illustrate the theory of §2.3 by the real quadratic field

$$k = \mathbb{Q}(\sqrt{5}) = \{\alpha = a + b\sqrt{5} \,|\, a, b \in \mathbb{Q}\}. \tag{2.25}$$

We determine those α's which are the roots of the equation

$$x^2 + Ax + B = 0, \ A, B \in \mathbb{Z} \tag{2.26}$$

and prove that these numbers (**algebraic integers** in k) form a ring (indeed, it is an integral domain but customarily called a ring) called the **ring of integers** in k and denoted by $\mathcal{O} = \mathcal{O}_k$.

Proof. If $\alpha = a + b\sqrt{5}$ is a root of (2.26), then the conjugate root is $\alpha' = a - b\sqrt{5}$, whose **norm** $N = N_{k/\mathbb{Q}}$ and **trace** $\text{Tr} = \text{Tr}_{k/\mathbb{Q}}$ being

$$N_{k/\mathbb{Q}} \, \alpha = a^2 - 5b^2, \quad \text{Tr}_{k/\mathbb{Q}} \, \alpha = 2a.$$

Since $2a = \text{Tr} \, \alpha = A$, we see that a is a half-integer: $a = \frac{a_1}{2}$, say, $a_1 \in \mathbb{Z}$. Hence the denominator of a^2 is at most 4 and so, in view of the fact that $a^2 - 5b^2 \in \mathbb{Z}$, that of $5b^2$ must be at most 4, or $b = \frac{b_1}{2}$, $b_1 \in \mathbb{Z}$. Hence it follows that $a^2 - 5b^2 = \frac{a_1{}^2 - 5b_1{}^2}{4}$ and $a_1{}^2 - 5b_1{}^2 \equiv a_1{}^2 - b_1{}^2 \bmod 4$, whence $a_1 . b_1$ must be of the same parity.

Then it is easily checked that α is of the form $a + b\tau$, $a, b \in \mathbb{Z}$, where

$$\tau = \frac{1 + \sqrt{5}}{2}, \tag{2.27}$$

and b is even or odd according as $a_1 \equiv b_1 \equiv 0 \bmod 2$ or $a_1 \equiv b_1 \equiv 1 \bmod 2$, whence we conclude that

$$\mathcal{O}_k = \mathbb{Z} \oplus \tau\mathbb{Z}. \tag{2.28}$$

$\square$

Note that the proof used to derive (2.28) applies to any quadratic fields and gives the integral basis: If $k = \mathbb{Q}(\sqrt{d}) = \{a + b\sqrt{d}, a, b \in \mathbb{Q}\}$, where d is a square-free integer, then $\mathfrak{O}$ is of the form (1.9), with τ in the form (2.27) $\frac{1 + \sqrt{d}}{2}$ if $d \equiv 1 \pmod 4$, while it is in the form $1 + \sqrt{d}$ if $d \equiv 2 \text{ or } 3 \pmod 4$. The discriminant D_k is d or $4d$, respectively.

The invertible elements of $\mathcal{O}$ are called **units** which constitute a group $\mathcal{O}^\times$ (**unit group**) and denoted by $U = U_k$. The following is known and easily proved (cf. e.g. [HaWr]).

- If $\xi = x + y\tau$, $x, y \in \mathbb{Z}$, then $N\xi = x^2 + xy - y^2$, an **indefinite quadratic form**.

- $\xi \in U$ if and only if $|N\xi| = 1$.

- The only elements of U of finite order are ± 1, which constitute the **torsion subgroup** $< -1 >$, with $< \alpha >$ designating the subgroup generated by α.

- A generator (> 1) of the group $U/\{\pm 1\}$ is called a **fundamental unit**.

In our case, τ is a fundamental unit and

Theorem 2.11. (Dirichlet unit theorem) $U = \{\pm 1\} \times < \tau >$.

Thus the field $\mathbb{Q}(\sqrt{5})$ is sometimes called the **field of the golden section unit**.

Theorem 2.12. $\mathbb{Q}(\sqrt{5})$ *is a Euclidean field and hence is a UFD, or what amounts to the same thing, the* **class number** $h = h_k$, *the order of the* **ideal class group** $C = I/P$ *is* 1.

2.6   Cyclotomic fields

Let $\eta = \eta_k = e^{2\pi i \frac{1}{k}}$ denote the first primitive k-th root of 1 in $\mathbb{C}$ and let μ_k denote the group of all k-th roots of 1, which is generated by η, We make an important remark that *the set of all k-th roots of 1 and the set of all primitive d-th roots of 1 for all $d|k$ are the same*. This fact is often described as

$$X^k - 1 = \prod_{d|k} \Phi_d(X), \tag{2.29}$$

where $\Phi_d(X) = \prod_{(m,k)=1}(X - \eta^m)$ is the d-th **cyclotomic polynomial** whose roots are all the primitive d-th roots of 1. (2.29) implies in particular (1.27).

We start by stating other equivalent formulations of $\varphi(q)$.

Recalling the cyclic group μ_q of q-th roots of 1 in Example 1.1, we have

$$\varphi(q) = \text{the number of } \textbf{primitive } q\textbf{-th roots of 1} \tag{2.30}$$

$$\text{(i.e. the generators) of } \mu_q.$$

Secondly, let ζ be a primitive q-th root of 1 and let $\mathbb{Q}(\zeta)$ be the q-th **cyclotomic field**. Since

$$\text{Gal}(\mathbb{Q}(\zeta)/\mathbb{Q}) \cong (\mathbb{Z}/q\mathbb{Z})^{\times}, \tag{2.31}$$

it follows that

$$\varphi(q) = |\text{Gal}(\mathbb{Q}(\zeta)/\mathbb{Q})| \tag{2.32}$$

and from Galois theory

$$\varphi(q) = [\mathbb{Q}(\zeta) : \mathbb{Q}], \tag{2.33}$$

the extension degree of $\mathbb{Q}(\zeta)$ over $\mathbb{Q}$.

Now we consider some interesting extensions of $\mathbb{Q}(\sqrt{5})$, which turn out to be of degree 4. First

Proposition 2.3. *The splitting field B of the polynomial $f(X) = X^4 - 8X^2 + 36$ is the biquadratic field $\mathbb{Q}(i, \sqrt{5})$, which is a Galois extension of $\mathbb{Q}$ with Galois group V_4, the Klein four group, i.e. the direct product of two cyclic groups of order 2 (recall Exercise 11). All the subfields of B (save for B and $\mathbb{Q}$) are $\mathbb{Q}(\sqrt{5})$, $\mathbb{Q}_4 = \mathbb{Q}(\sqrt{-4})$ and $k = \mathbb{Q}(\sqrt{-5})$.*

Proof.   The quadratic equation $X^2 - 8X + 36 = 0$ has the roots $4 \pm 2\sqrt{5}i$. Since $\sqrt{4 \pm 2\sqrt{5}i} = i \pm \sqrt{5}$, the roots of $f(X)$ are $\pm(i \pm \sqrt{5})$. Since these are contained in B and

$$\frac{1}{2}\left(i + \sqrt{5} - (i - \sqrt{5})\right) = \sqrt{5}, \ \frac{1}{2}(i + \sqrt{5} + i - \sqrt{5}) = i,$$

we have $B = \mathbb{Q}(i, \sqrt{5})$. Since B is obtained as the adjoint of i to $k = \mathbb{Q}(\sqrt{5})$ and $X^2 + 1$ is irreducible over k, B is of degree 4 over $\mathbb{Q}$. As the splitting field of the irreducible polynomial f, B is a Galois extension whose Galois group $G = \mathrm{Gal}(K/\mathbb{Q})$ is not cyclic. Since the groups of order 4 are Abelian, it must be isomorphic to V_4 by Exercise 11. It is given as $< \sigma > \times < \tau >$, where

$$\sigma(i) = -i, \sigma(\sqrt{5}) = \sqrt{5}, \ \tau(i) = i, \tau(\sqrt{5}) = -\sqrt{5},$$

and $\sigma^2 = 1, \tau^2 = 1, \sigma\tau = \tau\sigma$: $G = \{1, \sigma, \tau, \sigma\tau\}$. All the three subgroups of G and their fixed fields are given by

- $\{1, \sigma\}$ whose fixed field is $\mathbb{Q}(\sqrt{5})$,

- $\{1, \tau\}$ whose fixed field is $\mathbb{Q}(\sqrt{i})$,

- $\{1, \sigma\tau\}$ whose fixed field is $\mathbb{Q}(\sqrt{-5})$.   □

We use the following results from algebraic number theory to determine the class number $h = h_k$ and the Hilbert class field of $k = \mathbb{Q}(\sqrt{-5})$.

Theorem 2.13. *Let k be an arbitrary ground field (a finite extension of $\mathbb{Q}$) and let $\tilde{k}$ be its **Hilbert class field** (or **absolute class field**) of k, i.e. the maximal unramified Abelian extension of k. Then*

$$\mathrm{Gal}(\tilde{k}/k) \approx C = I/P. \tag{2.34}$$

By $\mathbb{Q}_m$ we mean the m-th **cyclotomic field** $\mathbb{Q}(\zeta_m)$ with ζ_m designating the primitive m-th root of 1 and $\mathbb{Q}(\zeta_m)_+ = \mathbb{Q}(\zeta_m + \zeta_m^{-1})$ its **maximal real subfield**, i.e. the real subfield k_+ with extension degree $[\mathbb{Q}(\zeta_m) : k_+] = 2$.

Theorem 2.14. (Kronecker-Weber theorem) *If $k/\mathbb{Q}$ is Abelian, then there exists an integer $2 < m \in \mathbb{N}$ such that*

$$k \subset \mathbb{Q}_m,$$

m being called an admissible modulus.

Corollary 2.2. *For an Abelian extension $k/\mathbb{Q}$, there exists the least admissible modulus $f = f_k$, called the **conductor**, such that $k \subset \mathbb{Q}_m$.*

Proposition 2.4. *If $k = \mathbb{Q}_m$ or $k = \mathbb{Q}_m^+ = \mathbb{Q}(\zeta_m + \zeta_m^{-1})$, then $f_k = m$. If $k = \mathbb{Q}(\sqrt{d})$ with d square-free, then*

$$f_k = \begin{cases} |d| & d \equiv 1 \bmod 4, \\ |4d| & d \equiv 2, 3 \bmod 4. \end{cases}$$

It follows that for $k = \mathbb{Q}(\sqrt{5})$, $f_k = 5$ and $k = \mathbb{Q}_5^+$, the maximal real subfield of the 5-th cyclotomic field. Since the Galois group $\mathrm{Gal}(\mathbb{Q}_5/\mathbb{Q})$ is isomorphic to $(\mathbb{Z}/5\mathbb{Z})^\times$ which is a cyclic group of order $\varphi(5) = 4$ and that of B is not cyclic, $k = \mathbb{Q}(\sqrt{5})$ and B are not the same. Cf. Exercise 41 below.

Theorem 2.15. (Conductor-ramification theorem) *Suppose $k/\mathbb{Q}$ be Abelian. Then the rational prime p ramifies in k if and only if $p|f_k$.*

The following formula is due to Dirichlet (cf. [Dick, p. 346]) and proved in a general form by Funakura [Funa]. We give a proof of a more general Theorem 5.11 in Chapter 5.

Proposition 2.5. *Let $k = \mathbb{Q}(\sqrt{d})$, $d < -2$ and suppose the conductor f_k of k is odd. Then*

$$h_k = \frac{1}{2 - \left(\frac{d}{2}\right)} \sum_{\substack{0 < a < f_k/2 \\ (a, f_k) = 1}} \left(\frac{d}{a}\right), \tag{2.35}$$

where $\left(\frac{d}{a}\right)$ is the Kronekcer symbol.

Example 2.4. For $k = \mathbb{Q}(\sqrt{-5})$ we have $h_k = 2$. Hence $\mathcal{O}_k$ is not a UFD. An example is $6 = 2 \cdot 3 = (1 + \sqrt{5})(1 - \sqrt{5})$. We also have $\tilde{k} = \mathbb{Q}(i, \sqrt{5})$.

Proof.   Proposition 5.2 reads in this case

$$h_k = \frac{1}{2 - \chi_{-4}(2)} \sum_{a=1,3,7,9} \chi_{-4}(a) \left(\frac{a}{5}\right), \qquad (2.36)$$

where χ_{-4} is the real primitive odd character to the modulus 4: $\chi_{-4}(-1) = -1$. Hence

$$h_k = \frac{1}{2} \left(1 + \chi_{-4}(-1) \left(\frac{-2}{5}\right) + \chi_{-4}(-1) \left(\frac{2}{5}\right) + \chi_{-4}(1) \left(\frac{2}{5}\right)\right) = 2 \qquad (2.37)$$

because $\left(\frac{-1}{5}\right) = (-1)^{\frac{5-1}{2}} = 1$ and $\left(\frac{2}{5}\right) = (-1)^{\frac{5^2-1}{8}} = -1$.

Hence by Theorem 2.13, $|\mathrm{Gal}(\tilde{k}/k)| = 2$.

By Galois theory, this implies $[\tilde{k} : k] = 2$. Since $\mathbb{Q}(i, \sqrt{5})/k$ is unramified, we have $\mathbb{Q}(i, \sqrt{5}) \subset \tilde{k}$, whence we conclude the second assertion. □

Exercise 41. Let p be an odd prime and let $\zeta_p = e^{\frac{2\pi i}{p}}$ be the first primitive pth root of 1. Prove that the cyclotomic field $\mathbb{Q}(\zeta_p)$ contains a unique quadratic subfield given either by $\mathbb{Q}(\sqrt{p})$ for $p \equiv 1 \,(\mathrm{mod}\,4)$ or $\mathbb{Q}(\sqrt{-4p})$ for $p \equiv -1 \,(\mathrm{mod}\,4)$.

Solution.   We use some well-known facts on cyclotomic fields. The Galois group of $\mathbb{Q}(\zeta_p)/\mathbb{Q}$ is isomorphic to $(\mathbb{Z}/p\mathbb{Z})^\times$ which is a cyclic group of order $\varphi(p) = p - 1$.

Now all the subgroups of a cyclic group are cyclic and they are determined uniquely by their orders which are the divisors of $p - 1$. Hence for each divisor d of $p - 1$, there is a unique (up to isomorphism) subgroup with order d, and *a fortiori* there is a unique subgroup of order $\frac{\varphi(p)}{2} = \frac{p-1}{2}$ to which there corresponds, by Galois theory, a unique fixed field of degree 2. Since the possible discriminants in our case are p or $-4p$ according as $p \equiv 1 \,(\mathrm{mod}\,4)$ or $p \equiv -1 \,(\mathrm{mod}\,4)$ (cf. e.g. [Dave, p. 43]), the assertion follows.

Exercise 42. Solve the 5-th cyclotomic equation $X^4 + X^3 + X^2 + X^2 + X + 1 = 0$ by two different ways.

To sum up, we have obtained

Theorem 2.16.   *The quadratic field of the golden section unit $k = \mathbb{Q}(\sqrt{5})$ is*

- *one of known real quadratic Euclidean fields.*

- *the maximal real subfield (as well as the unique real quadratic subfield) of the 5-th cyclotomic field $\mathbb{Q}_5 = \mathbb{Q}(\zeta_5)$.*

- *the unique real quadratic subfield of the bicyclic biquadratic field B which is the Hilbert class of k.*

Toward the end of this section, we shall make use of (2.29) to prove a special case of the celebrated **Dirichlet prime number theorem** to the effect that any arithmetic progression a mod q, $(a, q) = 1$ contains infinitely many primes. We remark that in [Waid] and interesting interpretation of Dirichlets theorem in terms of non-standard analysis.

Theorem 2.17. (R. A. Smith [Smit]) *The congruence*

$$X^q \equiv 1 \ (\mathrm{mod}\, p) \tag{2.38}$$

is solvable with an integer of order q mod p for infinitely many primes p, and a fortiori, there exist infinitely many primes p such that $p \equiv 1(\mathrm{mod}\, q)$.

For a proof we need a lemma.

Lemma 2.6. (T. Nagel) *If $f \in \mathbb{Z}[X]$ is not a constant, then the congruence*

$$f(x) \equiv 1 \ (\mathrm{mod}\, p) \tag{2.39}$$

is solvable for infinitely many primes p.

Proof. We prove this in the form:
$f(x)$ $(x \in \mathbb{Z})$ has infinitely many prime divisors.

Suppose $f(X) = a_0 X^m + \cdots + a_{m-1} + a_m$, $m \in \mathbb{N}$. If $a_m = 0$, then all the primes are prime divisors of $f(x)$. Hence we may suppose $a_m \neq 0$. Suppose the contrary to the assertion, i.e. that $f(x)$ has only finitely many prime divisors $p_1, \cdots, p_r$, say. Let $b = p_1 \cdots p_r a_m$. Then clearly,

$$f(bY) = a_m g(Y),$$

where $g(Y) = A_m Y^m + \cdots + A_1 Y + 1$, and where all A_i's are divisible by $p_1 \cdots p_r$.

Every prime divisor of $g(y)$ is a prime divisor of $f(x)$ and none of $p_1, \cdots, p_r$ can be a prime divisor of $g(y)$ in view of its constant term 1. Hence $g(y) = \pm 1$ for all $y \in \mathbb{Z}$. However, this equation $g(y) = \pm 1$ has at most $2m$ roots, and so we may find an integer y such that $g(y)$ is divisible by a prime p_{r+1} other than $g(y) = \pm 1$. Since p_{r+1} is a prime divisor of $f(x)$, we have a contradiction, completing the proof. $\square$

Proof of Theorem 2.17. Recall the p-adic valuation in Example 2.3: For $x \in \mathbb{Z}$, $x = p^{v_p(x)}y$ with $y \in \mathbb{Z}$, $(y,p) = 1$. Substituting the value of $x \in \mathbb{Z}$, $x \neq \pm 1$ in (2.29) and taking the p-adic valuation of both sides, we obtain

$$v_p(x^n - 1) = \sum_{d|n} v_p(\Phi_p(x)), \qquad (2.40)$$

which is inverted to

$$v_p(\Phi_n(x)) = \sum_{d|n} \mu\left(\frac{n}{d}\right) v_p(x^n - 1) \qquad (2.41)$$

by the Möbius inversion (3.4) (Chapter 3). By Lemma 2.6, there exist infinitely many primes p such that

$$\Phi_n(x) \equiv 1 \pmod{p} \qquad (2.42)$$

is solvable in $\mathbb{Z}$. Fix such a prime p and choose an $x > n$ satisfying (2.42). Recall the notion of the order of an element of a group (Definition 1.6) and let $f = o(x)$, i.e. the order of x in $(\mathbb{Z}/p\mathbb{Z})^{\times}$. Then clearly $f \mid n$. It therefore suffices to prove that $f = n$.

If $f \nmid d$, then $v_p(x^d - 1) = 0$. Hence (2.41) becomes

$$v_p(\Phi_n(x)) = \sum_{\substack{d|n \\ f|d}} \mu\left(\frac{n}{d}\right) v_p(x^n - 1) = \sum_{d|m} \mu\left(\frac{m}{d}\right) v_p(x^d f - 1) \qquad (2.43)$$

on writing df for d, where $m = \frac{n}{f}$, which is relatively prime to p.

Now by Exercise below, each summand $v_p(x^d f - 1)$ amounts to $v_p(x^f - 1)$, which can therefore be factored out, whereby we obtain

$$v_p(\Phi_n(x)) = v_p(x^f - 1) \sum_{d|m} \mu\left(\frac{m}{d}\right). \qquad (2.44)$$

By (1.41), Chapter 1 ((3.3), Chapter 3), the right-hand side of (2.44) is 0 if $m > 1$. But since $v_p(\Phi_n(x)) \geq 1$, it follows that $m = 1$, amounting to $f = n$. This complete the proof.

Exercise 43. Prove for $(d,p) = 1$ that $v_p(x^d f - 1) = v_p(x^f - 1)$.

2.7 The dihedral group as a Galois group

Generalizing the situation in Proposition 2.3 slightly, consider the biquadratic field $\mathbb{Q}(i, \sqrt{p})$ for an od prime p, which is a Galois extension of $\mathbb{Q}$ with Galois group V_4 and its 3 subfields of $\mathbb{Q}(\sqrt{p})$, $\mathbb{Q}_4 = \mathbb{Q}(\sqrt{-4})$ and $\mathbb{Q}(\sqrt{-p})$. P. Chowla proves the following theorem.

Theorem 2.18. *Suppose $p \equiv 5$ mod 8 and let h and ε be the class number of $\mathbb{Q}(\sqrt{p})$ and let H be the class number of $\mathbb{Q}(\sqrt{-p})$. Further let $\left\{ \frac{v}{2} \right\}$ denote the least (positive) solution of the congruence $2 \left\{ \frac{v}{2} \right\} \equiv v$ mod p. Then*

$$\varepsilon^h = (-1)^m \prod_{v < \frac{p}{2}} 2 \cos \frac{2\pi}{p} \left\{ \frac{v}{2} \right\}, \qquad (2.45)$$

where v runs through all quadratic residues $< \frac{p}{2}$ and

$$m = \frac{1}{2} \left\{ \frac{p-1}{4} + \frac{1}{2} H \right\}. \qquad (2.46)$$

In this section we shall prove (following [KKPa]) that Theorem 2.18 does not give any relation between the class number of the real and imaginary quadratic fields $\mathbb{Q}(\sqrt{p})$, $\mathbb{Q}(\sqrt{-p})$ but is a restatement of the classical theorem of Dirichlet ([BShh, Theorem 2, p. 362])

Theorem 2.19. (Dirichlet) *If k is a real quadratic field with discriminant d, class number h and the fundamental unit $\varepsilon > 1$, then*

$$\varepsilon^h = \frac{\prod \sin \frac{\pi n}{d}}{\prod \sin \frac{\pi v}{d}}, \qquad (2.47)$$

where v and n run through those integers in $(0, \frac{p}{2})$ such that $\left(\frac{v}{d} \right) = 1$ and $\left(\frac{n}{d} \right) = -1$.

Corollary 2.3. *If $p \equiv 5$ mod 8, then Chowla's theorem (Theorem 2.18) is a reformulation of Dirichlet's theorem (Theorem 2.19) in conjunction with another formula of Dirichlet*

$$2 S_{1/4}(\chi_p) = H, \qquad (2.48)$$

where $S_{1/4}(\chi_p) = \sum_{0 < a < p/4} \left(\frac{a}{p} \right)$ with $\chi_p(a) = \left(\frac{a}{p} \right)$ the Legendre symbol.

Proof. Since $\left(\frac{2}{p} \right) = -1$, the non-residues are given in the form $2v$. Hence (2.47) may be written as

$$\varepsilon^h = \prod \frac{\sin \frac{2\pi v}{p}}{\sin \frac{\pi v}{p}} = \prod_{0 < v < p/2} 2 \cos \frac{\pi v}{p}. \qquad (2.49)$$

Noting that $\left\{\frac{v}{2}\right\}$ also means the least positive residue mod p of $2^{-1}v$, where 2^{-1} is the inverse of 2 mod p:

$$\left\{\frac{v}{2}\right\} = 2^{-1}v - p\left[\frac{2^{-1}r}{p}\right],$$

we may express $\left\{\frac{v}{2}\right\}$ more concretely as

$$\left\{\frac{v}{2}\right\} = \begin{cases} \frac{v}{2} & v \text{ even}, \\ \frac{v+p}{2} & v \text{ odd}. \end{cases} \tag{2.50}$$

Hence if we replace v in (2.49) by $\left\{\frac{v}{2}\right\}$, then those terms with odd v are negative and there are m' of them, where m' indicates the number of odd residues in $(0, \frac{p}{2})$. Hence it follows that

$$\varepsilon^h = (-1)^{m'} \prod_{\substack{0<v<p/2 \\ 2\nmid v}} 2\cos 2\pi \left\{\frac{v}{2}\right\}. \tag{2.51}$$

The remaining part remains the same as in [PCho] and we may prove by elementary argument that

$$m' = \frac{1}{2}\left\{\frac{p-1}{4} + \frac{1}{4}S_{1/4}\right\},$$

which is m in view of (2.49), completing the proof.   □

Chapter 3

Arithmetical functions and Stieltjes integrals

This chapter is devoted to the development of elementary analytic number theory, with emphasis on the use of Stieltjes integrals and Hilbert spaces. The special notations in this chapter are R and G indicating the ring of all arithmetic functions and the group of multiplicative functions, respectively. We state three generating functions for arithmetic functions: power series, Dirichlet series and Euler products.

3.1 Arithmetical functions and their algebraic structure

Definition 3.1. We call any function

$$f : \mathbb{N} \to \mathbb{C}$$

an **arithmetical function** (or arithmetic function, number-theoretic function). The domain is sometimes extended to $\mathbb{Z}$ and the range may be replaced by a general integral domain.

We write

$$R = \{f | f : \mathbb{N} \to \mathbb{C}\},$$

the set of all arithmetic functions.

For an extensive theory of arithmetic functions, we refer to [MacC], [Siva].

Exercise 44. In R define the sum and the product of f, g by

$$(f + g)(n) = f(n) + g(n),$$

$$(fg)(n) = (f \cdot g)(n) = f(n)g(n).$$

Then prove that $(R, +, \cdot)$ is a ring.

We introduce another product, the **Dirichlet convolution**, which is crucial in multiplicative number theory.

Definition 3.2. For $f, g \in R$, define

$$(f * g)(n) = \sum_{d|n} f\left(\frac{n}{d}\right) g(d) = \sum_{d\delta = n} f(d)g(\delta), \qquad (3.1)$$

when d runs through all positive divisors of n.

Theorem 3.1. $(R, +, *)$ *is an integral domain with the unity* e *where*

$$e(n) = \begin{cases} 1, & n = 1 \\ 0, & n \neq 1. \end{cases}$$

$f \in R^\times$ *if and only if* $f(1) \neq 0$.

Proof. Commutative law and distributive law are clear. The associative law follows from

$$(f_1 * f_2) * f_3(n) = \sum_{d_1 d_2 d_3 = n} f_1(d_1) f_2(d_2) f_3(d_3).$$

Hence $(R, +, *)$ is a ring with the zero element as the zero map $0(n) = 0$.

By definition, $(e * f)(n) = \sum\limits_{d_1 d_2 = n} e(d_1) f(d_2) = f(n)$.

Now suppose that $f * g = 0, f \neq 0, g \neq 0$. Then there exist $\min\{n \in \mathbb{N} | f(n) \neq 0\} := k$, $\min\{n \in \mathbb{N} | g(n) \neq 0\} := l$. Hence

$$0 = (f * g)(kl) = \sum_{d\delta = kl} f(d)g(\delta).$$

The sum consists of three parts $d\delta = kl$, $d < k$; $d\delta = kl$, $\delta < k$, ; $d = k, \delta = l$. The first two parts are 0 and

$$0 = (f * g)(kl) = f(k)g(l) \neq 0,$$

a contradiction. Hence R is an integral domain.

Finally, suppose $f(1) \neq 0$, then we may define g satisfying $f * g = e$ inductively as:

$$g(1) = f(1)^{-1}, \quad g(n) = -f(1)^{-1} \sum_{d\delta = n, d \neq 1} f(d)g(\delta).$$

$\square$

For UFD property of R, cf. [CPLu].

Exercise 45. Let R' be the set of all arithmetical functions $f : \mathbb{N} \cup \{0\} \to \mathbb{C}$. Define the Abel convolution $f \times g$ by

$$(f \times g)(n) = \sum_{k=0}^{n} f(k)g(n-k) = \sum_{a+b=n} f(a)g(b).$$

Then prove that $R = (R, +, \times)$ is an integral domain with the unity e',

$$e'(n) = \begin{cases} 1, & n = 0 \\ 0, & n > 0. \end{cases}$$

As one can easily perceive, the genesis of the **Abel convolution** is the Cauchy product of power series. Suppose

$$F_1(z) = \sum_{n=0}^{\infty} f(n)z^n, \ G_1(z) = \sum_{n=0}^{\infty} g(n)z^n$$

be power series associated to f and g, respectively, which we assume are absolutely convergent in a disc $|z| < r, 0 < r$. Then the product $F_1(z)G_1(z)$ gives rise to

$$F_1(z)G_1(z) = \sum_{n=0}^{\infty} (f \times g)(n)z^n$$

in the same disc. Examples of sequences generated by power series will be given below (cf. e.g. Definition 3.5).

We now clarify the genesis of the Dirichlet convolution.

Definition 3.3. For an arithmetic function $f \in R$, we associate a Dirichlet series $F(s)$, called the **generating Dirichlet series**, which is to be convergent in some half-plane.

$$F(s) = \sum_{n=0}^{\infty} \frac{f(n)}{n^s}, \quad \sigma > \sigma_f,$$

where σ_f indicates the abscissa of absolute convergence.

Example 3.1. If

$$f(n) << n^{\alpha + \varepsilon}, \quad \forall \varepsilon > 0,$$

then σ_f can be taken $\sigma_f \leq \alpha$.

Given two arithmetical functions $f, g \in R$, we denote the associated generating Dirichlet series by $F(s)$ and $G(s)$, which are absolutely convergent for $\sigma > \max\{\sigma_f, \sigma_g\}$.

The product $F(s)G(s)$ also converges absolutely in the same half-plane, giving rise to $f * g$:

$$F(s)G(s) = \sum_{n=0}^{\infty} \frac{(f * g)(n)}{n^s}, \quad \sigma > \max\{\sigma_f, \sigma_g\}. \tag{3.2}$$

Formula (3.2) gives the genesis of (3.1) and these two are essential ingredients in the study of multiplicative properties of integers. (3.1) is finite and is sometimes useful to give information, while (3.2) is, though infinite, often more powerful than (3.1).

Many familiar arithmetic functions arise in this way from basic case, which we now introduce:

$$I(n) = 1, \quad n \in \mathbb{N} \longleftrightarrow F(s) = \zeta(s), \quad \sigma > 0;$$

$$N(n) = n, n \in \mathbb{N} \longleftrightarrow F(s) = \zeta(s-1), \quad \sigma > 2,$$

where $\zeta(s)$ indicates the Riemann zeta-function to be introduced by (3.69) below.

The Möbius function $\mu(n)$ is introduced as the coefficients of $\zeta(s)^{-1}$ (cf. (3.7)), where $\zeta(s)$ is introduced by (3.69):

$$\mu * I = e \longleftrightarrow \zeta(s)\zeta(s)^{-1} = 1$$

or

$$\sum_{d|n} \mu(d) = \begin{cases} 1 & n = 1 \\ 0 & n > 1. \end{cases} \tag{3.3}$$

For a generalization of (3.3), cf. Theorem 3.7 below. Further examples are

$$I * I = d \longleftrightarrow \zeta(s)\zeta(s) = \zeta(s)^2 = \sum_{n=1}^{\infty} \frac{d(n)}{n^s}, \sigma > 1,$$

where $d(n) = \sum_{d|n} 1$ is the **divisor function** counting the number of divisors of n.

$$I * N^{\alpha} = \sigma_{\alpha} \longleftrightarrow \zeta(s)\zeta(s-a) = \sum_{n=1}^{\infty} \frac{\sigma_{\alpha}(n)}{n^s},$$

where $\sigma_{\alpha}(n) = \sum_{d|n} d^{\alpha}$ is the **sum-of-divisor function**. We often write

$$\sigma_1 = \sigma \quad (\sigma_0 = d).$$

(3.3) is the most fundamental and form the basis of the Möbius inversion formula. To provide a clearer view of the picture, we make the following definition.

Definition 3.4. If $0 \neq f \in R$ satisfies

$$f(mn) = f(m)f(n) \quad (m, n) = 1, \tag{3.4}$$

then $f \in R$ is called a **multiplicative function**. If f satisfies

$$f(mn) = f(m) + f(n) \quad (m, n) = 1, \tag{3.4$'$}$$

then f is called an **additive function**. If (3.4) or (3.4)$'$ holds for all m, n, then f is called **completely multiplicative** (or **additive**).

Exercise 46. Prove the following

- f is additive, $c > 0 \Longleftrightarrow g(n) = c^{f(n)}$ is multiplicative;
- f is multiplicative, $f(n) > 0 \Longleftrightarrow g(n) = \log f(n)$ is additive.

The following theorem assures that the values of multiplicative and additive function are determined by those at prime powers.

Theorem 3.2. *Let* $n = \prod_p p^{\alpha_p(n)}$ *be the canonical decomposition. Then for a multiplicative function* f *and an additive function* g, *we have*

$$f(n) = \prod_p f(p^{\alpha_p(n)}), \quad f(1) = 1, \tag{3.5}$$

$$g(n) = \sum_p f(p^{\alpha_p(n)}), \quad g(0) = 0.$$

Proof. By (3.4),

$$f(m_1 m_2 \cdots m_n) = f(m_1)f(m_2) \cdots f(m_n),$$

if $m_1, m_2, \cdots, m_n$ are pairwise relatively prime. Applying this fact to $\{2^{\alpha_2(n)}, \cdots, p^{\alpha_p(n)}\}$, we conclude the desired expression. Since $f \neq 0$, there is an $n_0 \in X$ such that $0 \neq f(n_0) = f(n_0)f(1)$, whence $f(1) = 1$. Applying this fact to $f(n) := 2^{g(n)}$, then $2^{g(0)} = 1, g(0) = 0$.

Theorem 3.2 implies that if f is additive and $f(p^k) = 0$ (p all primes, $k \in \mathbb{N}$, then $f = 0$; and if f is completely additive, then we have $f(p^k) = kf(p)$, so $f(p) = 0$ (p all primes) implies $f = 0$. □

Example 3.2. The Euler function $\varphi(n) = (\mathbb{Z}/n\mathbb{Z})^\times$ ((1.26)) is multiplicative. This is because, for $(m, n) = 1$, $(\mathbb{Z}/mn\mathbb{Z})^\times = (\mathbb{Z}/m\mathbb{Z})^\times (\mathbb{Z}/m\mathbb{Z})^\times$ ((1.36)).

We shall dwell on other properties of $\varphi(n)$ below.

We now prove another algebraic structure theorem.

Theorem 3.3. *The set G of all multiplicative functions forms a group.*

We need a lemma which anticipates some knowledge of divisibility of integers.

Lemma 3.1. *If $(m, n) = 1$, then each $d|mn$ can be expressed uniquely in the form $d = d_1 d_2, d_1|m_1, d_2|n, (d_1, d_2) = 1$.*

Proof. Let $m = \prod_p p^{\alpha_p(m)}, n = \prod_p p^{\alpha_p(n)}$. Then since the primes dividing d are those which divide m or n, we have

$$d_1 = \prod_{p|m} p^{a_p}, \ d_2 = \prod_{p|m} p^{b_p}, \ d_1 d_2 = d,$$

where $a_p \le \alpha(m)$ and $b_p \le \alpha(n)$. Hence $d_1|m, d_2|n, (d_1, d_2) = 1$.

The uniqueness follows from the fact that d_1, respectively, d_2 is the largest divisor of m, respectively of n. □

Proof of Theorem 3.3. For $f, g \in G$, let $(m, n) = 1$. Since

$$(f * g)(mn) = \sum_{d\delta = mn} f(d)g(\delta),$$

we may express d, δ, by Lemma 3.1, as

$$d_1 d_2 = d, \ d_1|m, \ d_2|n, \ \delta = \delta_1 \delta_2, \ \delta_1|m, \delta_2|n.$$

Hence, noting that $d_1\delta_1 = m, d_2\delta_2 = n$,

$$(f * g)(mn) = \sum_{\substack{d_1|m,d_2|n \\ \delta_1|m,\delta_2|n}} f(d_1)g(\delta_1)f(d_2)g(\delta_2) = (f * g)(m)(f * g)(n).$$

$$(3.6)$$

That G forms a semi-group follows from this and Theorem 3.1.

Since $f(1) = 1 \ne 0$, by the same theorem, there exists a $g \in R^\times$ such that $f * g = e$.

We show that $g \in G$.

Suppose that $1 < k = mn, (m, n) = 1$ and inductively that for $1 \le r < k, r = ab, (a, b) = 1$ imply $g(r) = g(a)g(b)$. Then by Lemma 3.1,

$$0 = e(mn) = (f * g)(mn) = \sum_{d|mn} = \sum_{\substack{d_1|m,d_2|n \\ (d_1,d_2)=1}} f(d_1)f(d_2)g\left(\frac{mn}{d_1 d_2}\right).$$

The term corresponding to $d_1 = 1, d_2 = n$ is $g(mn)$ and other terms can be written as $f(d_1)f(d_2)g\left(\frac{m}{d_1}\right)g\left(\frac{n}{d_2}\right)$ by hypothesis.

On the other hand, by (3.6),

$$0 = (f * g)(m)(f * g)(n) = \sum_{d_1|m} f(d_1)g\left(\frac{m}{d_1}\right) \sum_{d_2|n} f(d_2)g\left(\frac{n}{d_2}\right)$$

in which the term corresponding to $d_1 = 1, d_2 = 1$ is $g(m)g(n)$, while other terms are $f(d_1)f(d_2)g\left(\frac{m}{d_1}\right)g\left(\frac{n}{d_2}\right)$.

Hence comparing these two expressions, we conclude that $g \in G$, completing the proof.

Corollary 3.1. *If $h = f * g \in G$, then either both $f, g \in G$, or neither of them belong to G.*

Proof.   If $f \in G$, then the inverse element $f^{-1} \in G$, by Theorem 3.3, we have

$$f * f^{-1} = e$$

whence $g = h * f^{-1}$ and $g \in G$.   □

Corollary 3.2. $\sigma_\alpha \in G$ *and in particular, $d \in G$ and $\sigma \in G$.*

Proof.   The proof follows from $\sigma_\alpha = I * N^\alpha$ and $I, N^\alpha \in G$.   □

Corollary 3.3. $\mu = I^{-1}$, *i.e. (3.3) holds.*

To give another proof of (3.3), it is necessary to know the values of μ, and for that purpose, the natural way is to appeal to the Euler product

$$\zeta(s)^{-1} = \prod_p (1 - p^{-s}) = \sum_{n=1}^{\infty} \frac{\mu(n)}{n^s}, \quad \sigma > 1 \qquad (3.7)$$

which is a consequence of the most important Euler product for the Riemann zeta function (also see (3.69))

$$\zeta(s) = \prod_p (1 - p^{-s})^{-1} = \sum_{n=1}^{\infty} \frac{1}{n^s}, \quad \sigma > 1.$$

The Euler product of (3.69) will be proved in Exercise 54 in §3.5 below.

Assuming the validity of (3.7) for the moment, we expand the product in (3.7) to find that

$$\mu(n) = \begin{cases} 1, & n = 1, \\ 0, & p^2|n, \\ (-1)^k, & n = p_1 p_2 \cdots p_k, p_i \text{ distinct,} \end{cases} \qquad (3.8)$$

which is a usual definition of μ given in most textbooks.

From (3.8) it is clear that $\mu \in G$.

Proof of Corollary 3.3. By Theorem 3.3, it suffices to check the values at prime powers p^k:

$$(\mu * I)(p^k) = \mu(1) + \mu(p) + \cdots + \mu(p^k) = 1 - 1 = 0.$$

Corollary 3.4. (Möbius inversion formula) *For $f, g \in R$, the following are equivalent:*

$$(i) \qquad I * f = g, \quad \sum_{d|n} f(d) = g(n), \quad n \in \mathbb{N};$$

$$(ii) \qquad f = \mu * g, \quad f(n) = \sum_{d|n} \mu(d) g\left(\frac{n}{d}\right), \quad n \in \mathbb{N};$$

$$(iii) \qquad \sum_{n \leq x} g(n) = \sum_{d \leq x} f(d) \left[\frac{x}{d}\right], \quad 1 \leq r \in \mathbb{R},$$

where $[x]$ indicates the greatest integer $\leq x$.

Proof.   That (i) and (ii) are equivalent is contained on Corollary 3.3.

(i)$\Longrightarrow$ (iii). Substituting the expression for g, we find that

$$\sum_{n \leq x} g(n) = \sum_{n \leq x} \sum_{d|n} f(d) = \sum_{d\delta \leq x} f(d) = \sum_{d \leq x} f(d) \sum_{\delta \leq \frac{x}{d}} 1 \qquad (3.9)$$

whose innermost sum is $\left[\frac{x}{d}\right]$, completing the proof.

(iii)$\Longrightarrow$ (i). Reversing the above argument, we arrive at the first equality in (3.9):

$$\sum_{n \leq m} g(n) = \sum_{n \leq m} \sum_{d|n} f(d) \qquad (3.10)$$

where we put $x = m \in \mathbb{N}$. Substituting (3.10) with m replaced by $m - 1$, we get

$$g(m) = \sum_{d|m} f(d),$$

i.e. (i) follows. This completes the proof. $\square$

Let Δ denote the **difference operator**, $\Delta f(n) = f(n+1) - f(n)$. Note that $\Delta \log n = \log\left(1 + \frac{1}{n}\right) \to 0$. We may now state Erdös theorem ([PErd]), which asserts that this limiting condition characterizes the logarithm function among additive arithmetic functions.

Theorem 3.4. (Erdös) *If an additive arithmetic functions f satisfies the condition*

$$\lim_{n\to\infty} \Delta f(n) = 0, \tag{3.11}$$

then we must have

$$f(n) = c \log n \tag{3.12}$$

for some constant c.

Proof. It suffices to prove (3.12) for n a prime power, i.e.

$$f(p^k) = c \log p^k, \tag{3.13}$$

for all prime powers p^k. We fix p^k and prove that

$$\frac{g(n)}{\log n} = \frac{f(n)}{\log n} - \frac{f(p^k)}{\log p^k} \to 0 \tag{3.14}$$

as $n \to \infty$, where we set

$$g(n) = f(n) - \frac{f(p^k) \log n}{\log p^k}. \tag{3.15}$$

Since $\Delta g(n) = \Delta f(n) - \frac{f(p^k)}{\log p^k} \log \left(1 + \frac{1}{n}\right)$. Equation (3.12) for g also holds true by (3.12) for f. Further g vanishes at $n = p^k$: $g(p^k) = 0$.

We construct the strictly decreasing sequence $\{q_j\}$ of successive quotients of n divided by p^{kj}. By the Euclidean division,

$$q_j = p^k q_{j+1} + r_j, \quad 0 \le r_j < p^k, \tag{3.16}$$

starting from $j = 0$ with $n = q_0$, where $q_{j+1} = \left[\frac{q_j}{p^k}\right]$. Let r denote the greatest integer such that $p^{k(r-1)} \le n$. Then solving this inequality, we get $r \le \left[\frac{\log n}{\log p^k}\right] + 1$, with $[y]$ indicating the integral part of y, i.e. the greatest integer not exceeding y. Then $q_r < p^k$. From this sequence $q = \left[\frac{n}{p^k}\right]_{j=0}^r$ we construct a sequence all of whose terms are relatively prime to p^k or p by subtracting a fixed positive integer $a < p$ from the quotient; $n_j = q_j - a = \left[\frac{n_{j-1}}{p^k}\right] - a$ if $p \mid q_j$. Then by the way of construction we have

$$n_j = p^k n_{j+1} + r_j, \quad 0 \le r_j < (a+1)p^k, \quad j = 1, \cdots, r-1. \tag{3.17}$$

By the additivity of g and $\gcd(n_j, p^k) = 1$, we obtain

$$g(p^k n_{j+1}) = g(p^k) + g(n_{j+1}) = g(n_{j+1}),$$

by the vanishingness condition. Hence, noting that

$$g(n_j) = g(n_j) + g(n_{j+1}) - g(p^k n_{j+1}) = g(n_{j+1}) + g(n_j) - g(p^k n_j)$$

and that

$$g(n_j) - g(p^k n_{j+1}) = g(n_j) - g(n_j - 1) + g(n_j - 1) - g(p^k n_{j+1}) \quad (3.18)$$

$$= \sum_{i=p^k n_{j+1}}^{n_j - 1} \Delta g(i),$$

we may express $g(n_j) - g(n_{j+1})$ as a telescoping series

$$g(n_j) - g(n_{j+1}) = \sum_{i=p^k n_{j+1}}^{n_j - 1} \Delta g(i). \quad (3.19)$$

By the same telescoping technique, we obtain

$$g(n) = g(n_0) = \sum_{j=0}^{r-1} (g(n_j) - g(n_{j+1})) + g(n_r), \quad (3.20)$$

whence substituting (3.19), we deduce that

$$g(n) = g(n_r) + \sum_{j=0}^{r-1} \sum_{i=p^k n_{j+1}}^{n_j - 1} \Delta g(i). \quad (3.21)$$

Now the double sum on the right of (3.21) may be written as $\sum_{k=1}^{N_r} \Delta g(m_k)$ with the increasing labels $\{m_k\}$, $m_1 = n_r p^k$, $m_{N_r} = n - 1$. In view of (3.12) and regularity of the $(C, 1)$-mean, it follows that

$$\lim_{n \to \infty} \frac{1}{N_r} \sum_{k=1}^{N_r} \Delta g(m_k) = 0. \quad (3.22)$$

Also the number N_r of terms is estimated by

$$\sum_{j=0}^{r-1} \sum_{i=p^k n_{j+1}}^{n_j - 1} 1 \le r \max_{0 \le j \le r-1} r_j \le \left(\frac{\log n}{\log p^k} + 1 \right) (a+1) p^k \le c \log n \quad (3.23)$$

with a constant $c > 0$, by (3.17) and the estimate on r.

It remains to estimate (3.21) divided by $\log n$, thereby we note that since $n_r < p^k$, $|g(n_r)| \le \max_{1 \le j \le p^k} |g(i)| =: C$, say. Hence it follows that

$$0 \le \frac{|g(n)|}{\log n} \le \frac{C}{\log n} + \frac{1}{N_r} \sum_{k=1}^{N_r} |\Delta g(m_k)| \to 0 \quad (3.24)$$

as $n \to \infty$, thereby proving (3.14). Hence it follows that $\lim_{n \to \infty} \frac{f(n)}{\log n} = c$, say, must exist and be equal to $\frac{f(p^k)}{\log p^k}$, i.e. (3.13) follows, completing the proof.   □

3.2 Asymptotic formulas for arithmetical functions

Recall the divisor function $d(n)$, which takes on the value 2 for $n = p$ a prime and we also have an estimate $d(n) << n^\varepsilon$ for every $\varepsilon > 0$. We can also determine the maximal order of $d(n)$. However, to look at the behavior of $d(n)$ at large, it is customary to consider the sum of their values up to x:

$$D(x) = \sum_{n \le x} d(n),$$

which is called the **summary function** of $d(n)$. The main objective is to find the main term with a plausibly good error estimate.

We recall the procedure which we adopted in the course of proof of (3.9). It is a discrete analogue of the technique of transforming a double integral into a respected integral:

$$\int\int_D f(x, y)\mathrm{d}x\mathrm{d}y = \int_a^b \mathrm{d}y \int_{\varphi_1(y)}^{\varphi_2(y)} f(x, y)\mathrm{d}x$$

if $D = \{(x, y)|a \le y \le b, \varphi_1(y) \le x \le \varphi_2(y)\}$.

In our case, the arithmetic function a is often given in the form $f * g$. Then the summary function $A(x) = \sum_{n \le x} a(n)$ is the double sum

$$A(x) = \sum_{d\delta \le x} f(d)g(\delta).$$

The above technique of "double into repeated" works when an asymptotic formula for $F(x) = \sum_{n \le x} f(n)$ is known and $g(n)$ is smaller in a certain series – one condition is that $g(p) = o(|f(p)|)$.

We give two illustrative examples.

The first example is Euler's function in Example 3.2. Since $\varphi \in G$, it suffices to know the values at prime powers p^α. Since $\varphi(p^\alpha)$ is the number of integers $\le p^\alpha$ prime to p^α, it is clear that

$$\varphi(p^\alpha) = p^\alpha - p^{\alpha-1}.$$

Consider the convolution $\mu * N \in G$, which takes on the value $\mu(1)p^\alpha + \mu(p)p^{\alpha-1} = p^\alpha - p^{\alpha-1}$. Hence $\varphi = \mu * N$:

$$\varphi(n) = (\mu * N)(n) = \sum_{d|n} \mu\left(\frac{n}{d}\right) d. \tag{3.25}$$

Example 3.3. We have the asymptotic formula

$$\Phi(n) = \sum_{n \leq x} \varphi(n) = \frac{1}{2\zeta(2)}x^2 + xE(x) + O(x), \qquad (3.26)$$

where

$$E(x) = -\sum_{n \leq x} \frac{\mu(n)}{n}\overline{B}_1\left(\frac{x}{n}\right) = O(\log x), \qquad (3.27)$$

with $\overline{B}_1(x)$ designating the first periodic Bernoulli polynomial $x - [x] - \frac{1}{2}$:

$$\overline{B}_1(x) = x - [x] - \frac{1}{2}. \qquad (3.28)$$

Proof of (3.26). Since

$$\Phi(x) = \sum_{mn \leq x} m\mu(n) = \sum_{n \leq x} \mu(n) \sum_{n \leq \frac{x}{n}} n,$$

and

$$\sum_{n \leq x} n = \frac{1}{2}[x]([x]+1) = \frac{1}{2}\left(x - \overline{B}_1(x) - \frac{1}{2}\right)\left(x - \overline{B}_1(x) + \frac{1}{2}\right)$$

$$= \frac{1}{2}x^2 - x\overline{B}_1(x) + O(1),$$

it follows that

$$\Phi(x) = \frac{x^2}{2}\sum_{n \leq x}\frac{\mu(n)}{n^2} - xE(x) + O(x).$$

Writing

$$\sum_{n \leq x}\frac{\mu(n)}{n^2} = \sum_{n=1}^{\infty}\frac{\mu(n)}{n^2} - \sum_{n > x}\frac{\mu(n)}{n^2} = \frac{1}{\zeta(x)} + O(x),$$

we conclude (3.26).

Example 3.4. Consider the divisor function $d(n) = I * I(n) = \sum_{d|n} d$. Then

$$D(x) = \sum_{n \leq x} d(n) = x \log x + O(x).$$

Indeed, $D(x) = \sum_{n \leq x}\left[\frac{x}{n}\right] = \sum_{n \leq x}\left(\frac{x}{n} - B_1\left(\frac{x}{n}\right) - \frac{1}{2}\right] = x\sum_{n \leq x}\frac{1}{n} + O(x).$

Using the asymptotic formula

$$\sum_{n \leq x}\frac{1}{n} = \log x + O(1),$$

we conclude the assertion.

Remark 3.1. (i) (3.3) appears in many other contexts. One important principle **relative primality principle** is that when we have a sum $S = S(x)$ over $n \leq x$, $(n, q) = 1$ for a given integer $q > 0$,

$$S = \sum_{\substack{n \leq x \\ (n,q)=1}} f(n), \tag{3.29}$$

then we may replace the relatively prime condition by $\sum_{d|(n,q)} \mu(d)$ and writing dn for n, we transform S into

$$S = \sum_{d|q} \mu(d) \sum_{n \leq x/d} f(dn). \tag{3.30}$$

For example, if $f \equiv I$, then since $\sum_{n \leq x/d} 1 = \frac{q}{d}$, we deduce (3.25) again.

Similarly, if $f \equiv N$, then $\sum_{\substack{n \leq x \\ (n,q)=1}} n = \frac{1}{2} \sum_{d|q} d\mu(d) \left[\frac{x}{d}\right] \left(\left[\frac{x}{d}\right] + 1\right)$.

Another example appears in the proof of (3.90), Exercises 58 and 79 in Chapter 5.

(ii) In Examples 3.3 and 3.4, we appealed to asymptotic formulas for the summatory function of powers of natural numbers $\sum_{n \leq x} n$, $\sum_{n \leq x} 1$, $\sum_{n \leq x} \frac{1}{n}$.

These are not isolated and can be unified in the following far-reaching theorem, (3.32) (cf. [Vist, Chapter 3]).

Theorem 3.5. (Integral Representations, [Vist, Theorem 3.1, p. 55]) *For any $l \in \mathbb{N}$ with $l > \mathrm{Re}\, u + 1$ and for any $x \geq 0$, we have the integral representation*

$$L_u(x, a) = \sum_{r=1}^{l} \frac{\Gamma(u+1)}{\Gamma(u+2-r)} \frac{(-1)^r}{r!} \overline{B_r}(x) \, (x+a)^{u-r+1}$$

$$+ \frac{(-1)^l}{l!} \frac{\Gamma(u+1)}{\Gamma(u+1-l)} \int_x^\infty \overline{B_l}(t) \, (t+a)^{u-l} \, dt \tag{3.31}$$

$$+ \begin{cases} \dfrac{1}{u+1} (x+a)^{u+1} + \zeta(-u, a), & u \neq -1 \\ \log(x+a) - \psi(a), & u = -1 \end{cases}$$

where $\Gamma(s)$ and $\psi(s)$ are the gamma function and Gauss' digamma function. Also the asymptotic formula

$$L_u(x, a) = \sum_{r=1}^{l} \frac{(-1)^r}{r} \binom{u}{r-1} \overline{B_r}(x)(x+a)^{u-r+1} + O\left(x^{\mathrm{Re}(u)-l}\right)$$

$$+ \begin{cases} \dfrac{1}{u+1} (x+a)^{u+1} + \zeta(-u, a), & u \neq -1 \\ \log(x+a) - \psi(a), & u = -1 \end{cases} \tag{3.32}$$

holds true as $x \to \infty$.

Furthermore, the integral representation

$$\zeta(-u, a) = a^u - \frac{1}{u+1} a^{u+1} - \sum_{r=1}^{l} \frac{(-1)^r}{r} \binom{u}{r-1} B_r \, a^{u-r+1}$$
$$+ (-1)^{l+1} \binom{u}{l} \int_0^\infty \overline{B_l}(t)(t+a)^{u-l} dt, \tag{3.33}$$

which follows from (3.31) *by putting* $x = 0$, *holds true for all complex* $u \neq -1$, *where* l *can be any natural number subject only to the condition that* $l > \operatorname{Re} u + 1$.

Here $\overline{B}_r(x) = B_r(x - [x])$ ((3.28) being the first one) and B_r are the r-th periodic Bernoulli polynomial and the r-th Bernoulli number respectively, to be introduced in §3.3 below, and where $B_r(x)$ is the r-th periodic Bernoulli polynomial.

3.3 Generating functionology

As already stated in Definition 3.3, the Dirichlet series generate arithmetic functions as their coefficients. As is referred to in the passage about the Abel convolution in §3.1, power series, being mostly Taylor series for certain known functions, work as generating functions of many important arithmetic functions which appear as Taylor coefficients. In this section we collect some important arithmetic functions generated by power series. Power series has been used much longer than Dirichlet series and there is a rich source of results obtained, especially, in the field of combinatorics (cf. e.g. [Comt]). As is well-known, the region of absolute convergence of power series is a circle of finite or infinite radius (of convergence), while that of Dirichlet series is a half-plane.

Exercise 47. Prove that the function
$$f(z) = \frac{1}{e^z - 1}$$
has a simple pole at $z = 0$ with residue 1, while the function $zf(z) = \frac{z}{e^z - 1}$ has a removable singularity at $z = 0$.

Solution. The first assertion follows from the Laurent series
$$\frac{1}{e^z - 1} = \frac{1}{z + \frac{z^2}{2!} + \frac{z^3}{3!} + \cdots} \tag{3.34}$$
$$= \frac{1}{z} + f_0(z),$$

where $f_0(z) = -\frac{1}{2} + \frac{1}{12}z + \cdots$ is analytic at $z = 0$. Since $e^z = 1$ occurs for $z = 2\pi in$, $n \in \mathbb{Z}$, the Laurent expansion (3.34) holds in the annulus $0 < |z| < 2\pi$.

The second assertion follows from the following argument. Since near $z = 0$, $zf(z) = 1 - \frac{1}{2}z + \frac{1}{12}z^2 + \cdots$, it follows that the new function

$$g(x) = \begin{cases} 1, & z = 0 \\ \frac{z}{e^z - 1}, & z \neq 0 \end{cases}$$

is analytic inside the circle $|z| < 2\pi$.

Definition 3.5. The product $\frac{z}{e^z-1}e^{xz}$ is analytic in $|z| < 2\pi$ and has a Taylor expansion of the form:

$$\frac{ze^{xz}}{e^z - 1} = \sum_{n=0}^{\infty} \frac{B_n(x)}{n!} z^n, \quad |z| < 2\pi. \tag{3.35}$$

The n-th coefficient $B_n(x)$ is called the n-th **Bernoulli polynomial**.

$B_n = B_n(0)$ is called the n-th **Bernoulli number**. They are therefore generated by the power series

$$\frac{z}{e^z - 1} = \sum_{n=0}^{\infty} \frac{B_n}{n!} z^n, \quad |z| < 2\pi, \tag{3.36}$$

whence

$$B_0 = 1, \ B_1 = -\frac{1}{2}, \ B_2 = \frac{1}{6}, \ B_4 = -\frac{1}{30}, \ B_{2k+1} = 0 \ (k = 1, 2, \cdots). \tag{3.37}$$

In the same way, we may define the **Euler polynomials** $E_n(x)$ by

$$\frac{2e^{xz}}{e^z + 1} = \sum_{n=0}^{\infty} \frac{E_n(x)}{n!} z^n, \quad |z| < \pi \tag{3.38}$$

the series being absolutely and uniformly convergent in $|z| < \pi$ (the singularity from the origin being $z = \pm i\pi$).

As stated on [Serr, p. 90, Footnote (2)], there are several definitions of Bernoulli numbers. The most commonly used ones are those in the *b*-notation in [Serr] while Leopoldt's definition differs only at one value B_1 which is defined to be $\frac{1}{2}$ rather than $-\frac{1}{2}$. Here we followed Washington's notation [Wash] and introduced the B_n by (3.36). For a systematic account of Bernoulli polynomials, we refer to [Vist, Chapter 1] and we freely use the results from it. There are many generalizations of Bernoulli numbers and polynomials, some of which will be given in Chapter 5. Most of them have been introduced so as to express the special values of the relevant zeta- and L-functions at negative integral argument, whence at certain positive integral arguments, while the Bernoulli numbers themselves were used to express the sum of powers of natural numbers up to n, say, which were used by Euler to solve the Basler problem $\zeta(2) = \frac{\pi^2}{6}$.

Finally we mention a fundamental convergence theorem for general Dirichlet series.

Definition 3.6. Let a_n be an arbitrary sequence of complex numbers, $\{\lambda_n\}$ be an increasing sequence of real numbers with $\lambda_1 \geq 1$ and let $\lambda_n^{-s} = e^{-\log \lambda_n}$ with log meaning the principal value. Then the series of the form $F(s) = \sum_{n=1}^{\infty} a_n \lambda_n^{-s}$ are called (general) **Dirichlet series**. From the tradition of Riemann, we write $s = \sigma + it$.

Theorem 3.6. *(i) If the series $F(s)$ is convergent at a point s_0, then it is convergent in the half-plane $\sigma > \sigma_0$ and is uniformly convergent in the angular domain $|\arg(s - s_0)| \leq \theta < \frac{\pi}{2}$ and represents an analytic function there.*
(ii) If the series $F(s)$ is absolutely convergent at a point s_0, then it is so in the half-plane $\sigma > \sigma_0$ (in which $F(s)$ is analytic).

Proof uses the partial summation formula. Cf. e.g. [Hard].

3.4   Hilbert space and number theory

The topic mentioned in the title has been developed independently by Romanoff [Roma] and Wintner [Wint]. We mostly present the results of Romanoff in a slightly modified form.

On the ring R of all arithmetical functions, we introduce some additional operations.

First
$$Sf = I * f,$$
where I is the constant function in §3.1, so that by Corollary 3.3, $S^{-1}f = \mu * I$:

$$Sf(n) = \sum_{d|n} f(d), \tag{3.39}$$

$$S^{-1}f(n) = \sum_{d|n} \mu\left(\frac{n}{d}\right) f(d). \tag{3.40}$$

Also $Ef = e * f$, where e is the unity in R (Theorem 3.1):

$$Ef(n) = f(n). \tag{3.41}$$

For a given integer N,

$$Af(n) = f((n, N)), \tag{3.42}$$

and

$$Bf(n) = \varepsilon\left(\frac{N}{n}\right) f(n), \tag{3.43}$$

where $\varepsilon(n) = 1$ if $n \in \mathbb{Z}$ and 0 otherwise. It is the characteristic function of $\mathbb{Z}$ in $\mathbb{R}$.

Exercise 48. Prove that Corollary 3.3 reads

$$SS^{-1} = S^{-1}S = E. \tag{3.44}$$

Exercise 49. As in Definition 3.3, let $f(s)$ denote the generating Dirichlet series for $f(n)$, absolutely convergent for $\sigma > \sigma_f$. Then show that (3.39) and (3.40) may be expressed as

$$\zeta(s)F(s) = \sum_{n=1}^{\infty} \frac{Sf(n)}{n^s}, \quad \sigma > \max\{\sigma_f, 1\} \tag{3.45}$$

and

$$\frac{1}{\zeta(s)}F(s) = \sum_{n=1}^{\infty} \frac{S^{-1}f(n)}{n^s}, \quad \sigma > \max\{\sigma_f, 1\}, \tag{3.46}$$

respectively.

Exercise 50. Prove that

$$SB = AS, \tag{3.47}$$

whence that

$$BS^{-1} = S^{-1}A. \tag{3.48}$$

Solution. By definition

$$SBf(n) = \sum_{d|n} \varepsilon\left(\frac{N}{d}\right) f(d)$$

and the summand of the right-side sum is non-zero only if $d|N$, so that d must divide both n and N, or $d|(n, N)$, i.e.

$$SBf(n) = \sum_{d|(n,N)} f(d) = ASf(n).$$

Hence (3.47) follows.

From (3.47) we have

$$B = S^{-1}AS \tag{3.49}$$

whence (3.48) follows.

We are in a position to prove the following theorem, which is a sort of generalization of (3.3).

Theorem 3.7. (Romanoff) *For any* $f \in R$ *and* $k < n$, *we have*

$$\sum_{d|n} \mu\left(\frac{n}{d}\right) f((d,k)) = 0. \tag{3.50}$$

Proof.   We have on using (3.49) on the way,

$$\sum_{d|n} \mu\left(\frac{n}{d}\right) f((d,N)) = S^{-1}f((n,N)) = S^{-1}Af(n) = BS^{-1}f(n). \tag{3.51}$$

Putting $N = k < n$, we conclude (3.50) by (3.51). $\qquad\qquad\square$

Exercise 51. Prove the relation

$$\sum_{d|n,(d,k)=\delta} \mu\left(\frac{n}{d}\right) = 0 \tag{3.52}$$

for each $\delta|k$, and

$$\sum_{d|n} \mu\left(\frac{n}{k}\right) f((d,k)) = \sum_{\delta} f(\delta) \sum_{d|n,(d,k)=\delta} \mu\left(\frac{n}{d}\right). \tag{3.53}$$

Theorem 3.7 follows also from (3.52) and (3.53).

Definition 3.7. (Romanoff ([Roma], [KaKi])) Let H be a (complex) Hilbert space with inner product denoted by (f,g). Any sequence $\{f_n\} \subset H$ is said to have a D-**property**, or more precisely, D_g-property, if $(f_m, f_n) = g((m,n))$ for a given arithmetical function $g \in R$.

Exercise 52. Let $\{f_n\} \subset H$ have the D_g-property, and define a new sequence $\{\phi_n\} \subset H$ by

$$\phi_n = \sum_{d|n} \mu\left(\frac{n}{d}\right) f_d, \tag{3.54}$$

where the right-hand side may be viewed as a Dirichlet convolution $\mu * f$ by abuse of language. Then prove that the $\{\phi_n\}$ is orthogonal.

Solution For $k < n$, we have

$$(\phi_n, f_k) = \left(\sum_{d|n} \mu\left(\frac{n}{d}\right) f_d, f_k \right) = \sum_{d|n} \mu\left(\frac{n}{d}\right)(f_d, f_k) \tag{3.55}$$

$$= \sum_{d|n} \mu\left(\frac{n}{d}\right) g((d, k)) = 0$$

by Theorem 3.7. Hence if $m < n$, then ϕ_m being a linear combination of $f_k, (k \leq m < n)$, we have $(\phi_m, \phi_n) = 0$. Similarly, for $m > n$, $(\phi_m, \phi_n) = \overline{(\phi_n, \phi_m)} = 0$. Hence the result follows.

We write

$$G(n) := (\mu * g)(n) = \sum_{d|n} \mu\left(\frac{n}{d}\right) g(d) \tag{3.56}$$

throughout in this section. Equation (3.56) is equivalent to

$$g(n) = \sum_{d|n} G(d). \tag{3.57}$$

Exercise 53. Use (3.55) and the definition (3.54) to establish that

$$(\phi_n, \phi_n) = G(n). \tag{3.58}$$

Solution By (3.55), we have

$$(\phi_n, \phi_n) = \sum_{d|n} \mu\left(\frac{n}{d}\right)(\phi_n, f_d) = (\phi_n, f_n).$$

Hence, substituting (3.54) again,

$$(\phi_n, \phi_n) = \sum_{d|n} \mu\left(\frac{n}{d}\right)(f_d, f_n) = \sum_{d|n} \mu\left(\frac{n}{d}\right) g((d, n)),$$

which is the right-hand side of (3.58).

The following theorem gives a condition for D_g-property.

Theorem 3.8. *A necessary and sufficient condition for a sequence to exist in H having the D_g-property is that $g(n)$ satisfy the condition*

$$G(n) = (\mu * g)(n) \geq 0, \quad n \in \mathbb{N}. \tag{3.59}$$

Proof. Necessity follows from (3.58) and $(\phi_n, \phi_n) \geq 0$.

Conversely, suppose $G(n) \geq 0$ and let $\{\varphi_n\}$ be an arbitrary ONS of H. Define a new sequence $\{f_n\}$ by

$$f_n = \sum_{d|n} \sqrt{G(d)}\varphi_d. \tag{3.60}$$

Then $\{f_n\}$ has the D-property. This is immediate since

$$(f_n, f_m) = \sum_{d|n} \sqrt{G(d)} \sum_{\delta|m} \sqrt{G(\delta)}(\varphi_d, \varphi_\delta) = \sum_{d|(n,m)} G(d) = g((n,m)),$$

by (3.57). $\hspace{1cm}\square$

Theorem 3.9. *Let $G(n)$ be an arbitrary arithmetical function with $G(n) > 0$ and define $g(n)$ by (3.57). If two sequences $\{f_n\}$ and $\{\varphi_n\}$ are related by (cf. (3.54))*

$$f_n = \sum_{d|n} \sqrt{G(d)}\varphi_d \quad n \in \mathbb{N}, \tag{3.61}$$

then $\{\varphi_n\}$ is an ONS if and only if $\{f_n\}$ has the D-property.

Proof. Suppose $\{f_n\}$ has the D-property. Then, since (3.61) amounts to

$$\varphi_n = \frac{1}{\sqrt{G(d)}} \sum_{d|n} f_d = \frac{1}{\sqrt{G(d)}}\phi_n, \tag{3.62}$$

with ϕ_n in (3.54). Since by Exercise 52, $\{\phi_n\}$ is orthogonal and by (3.58), $\{\varphi_n\}$ is a normalization of $\{\phi_n\}$, it follows that $\{\varphi_n\}$ is an ONS.

Conversely, if $\{\varphi_n\}$ is an ONS, then as in the proof of Theorem 3.8, $\{f_n\}$ has a D-property. $\hspace{1cm}\square$

Theorem 3.10. *Let $\omega(n)$ be a completely multiplicative function, which takes non-zero values and square summable: $\sigma := \sum_{k=1}^{\infty} |\omega(k)|^2 < \infty$ and let α_n be an ONS in H. Define f_n by*

$$f_n = \overline{\omega(n)^{-1}} \sum_{k=1}^{\infty} \omega(k)\alpha_{nk}, \quad n \in \mathbb{N}. \tag{3.63}$$

Then f_n has the D_g-property with

$$g(n) = \frac{\tilde{\sigma}}{|\omega(n)|^2}. \tag{3.64}$$

Proof. Noting that

$$f_n = \overline{\omega(n)^{-1}} \sum_{k \equiv 0 \,(\mathrm{mod}\, n)} \omega\left(\frac{k}{n}\right)\alpha_k$$

and using the continuity of the inner product, we find that

$$(f_n, f_m) = \overline{\omega(n)^{-1}}\omega(m)^{-1} \sum_{k \equiv 0 \,(\mathrm{mod}\, n)} \sum_{\ell \equiv 0 \,(\mathrm{mod}\, m)} \omega\left(\frac{k}{n}\right)\bar{\omega}\left(\frac{\ell}{m}\right)(\alpha_k, \alpha_\ell),$$

$$\tag{3.65}$$

whose right-hand side is

$$\overline{\omega(n)^{-1}}\omega(m)^{-1} \sum_{k \equiv 0 \,(\text{mod}\,[n,m])} \omega\left(\frac{k}{n}\right)\bar{\omega}\left(\frac{k}{m}\right)$$

$$= |\omega(n)|^{-2}|\omega(m)|^{-2} \sum_{k \equiv 0 \,(\text{mod}\,[n,m])} |\omega(k)|^2,$$

with $[n, m]$ denoting the l.c.m. of n and m. Factoring $|\omega([n, m])|^2$ out and using the well-known relation $mn = [n, m](n, m)$, we conclude that

$$(f_n, f_m) = \frac{\tilde{\sigma}}{|\omega((n, m))|^2}, \tag{3.66}$$

which is $g((n, m))$ in (3.64), completing the proof. □

We are now in a position to state the main result of this section.

Theorem 3.11. *Notation being the same as in Theorem 3.10, define $\varphi(n)$ as in Theorem 3.9:*

$$\varphi_n = (\tilde{\sigma}\Omega(n))^{-1/2} \sum_{d|n} \mu\left(\frac{n}{d}\right) \overline{\omega(d)^{-1}} \sum_{k=1}^{\infty} \omega(k)\alpha_{dk}, \quad n \in \mathbb{N}. \tag{3.67}$$

Then φ_n is an ONS, where

$$\Omega(n) = \sum_{d|n} \mu\left(\frac{n}{d}\right) |\omega(d)|^{-2} = |\omega(n)|^{-2} \prod_{p|n} \left(1 - |\omega(p)|^2\right), \tag{3.68}$$

with p ranging over all prime divisors of n.

Many identities can be derived using Theorem 3.11. A typical example of the choice of g and Ω is $\omega(n) = n^{-s}$ with $\sigma = \text{Re}\, s > \frac{1}{2}$ and then $\Omega(n) = n^{2\sigma} \prod_{p|n} \left(1 - p^{-2\sigma}\right)$ and $\tilde{\sigma} = \zeta(2\sigma)$.

The last equality in (3.68) is the Euler product whose theory will be given in the next section.

3.5   Euler products

In his epoch-making paper, the unique paper on number theory in his whole career, Riemann [Rie1, p. 144] introduces $\zeta(s)$ in the form

$$\zeta(s) = \prod_{p}(1 - p^{-s})^{-1} = \sum_{n=1}^{\infty} \frac{1}{n^s} \tag{3.69}$$

and NOT the additive form first. This suggests that Riemann was well aware of the importance of the Euler product, which is one the main contributions of Euler. However, to our best knowledge at present, we cannot treat the product well and must be satisfied with the additive form. It is to be remarked that many of the most advanced zeta-functions including those of Artin, Selberg etc. are introduced in the form of Euler products, which are then expanded into Dirichlet series. As can be seen from the proof below, (3.69) is analytic equivalent of the unique factorization property of the rational integers.

Exercise 54. Give a proof of (3.69) by considering the finite product

$$\prod_{p \leq x} (1 - p^{-s})^{-1}, \quad \sigma > 1. \tag{3.70}$$

Solution. Expanding $(1 - p^{-s})^{-1}$ into the binomial series $\sum_{m=0}^{\infty} p^{-ms}$ and multiplying out, we deduce that

$$\prod_{p \leq x} (1 - p^{-s})^{-1} = \prod_{p \leq x} \sum_{k=0}^{\infty} p^{-ks} = {\sum_{n}}' n^{-s}, \tag{3.71}$$

where the last sum extends over the set A_x all natural numbers whose prime factors are $\leq x$, where of course, we appeal to $\mathbb{Z}$ being a UFD. Hence

$$\left| \zeta(s) - \prod_{p \leq x} (1 - p^{-s})^{-1} \right| \leq \sum_{n \geq x} n^{-\sigma} = O\left(x^{1-\sigma}\right) \to 0$$

as $x \to \infty$, whence (3.69) follows.

Theorem 3.12. *If $f(n)$ is multiplicative and the series $S = \sum_{n=1}^{\infty} f(n)$ is absolutely convergent, then so is the infinite product $P = \prod \left(\sum_{k=0}^{\infty} f(p^k) \right)$ and $S = P$ holds, where p runs through all the primes.*

Proof of this theorem is verbatim to the solution to Exercise 54. Indeed, (3.71) should read

$$\prod_{p \leq x} \sum_{k=0}^{\infty} f(p^{-k}) = {\sum_{n \in A_x}}' f(n),$$

and the subsequent equation should read

$$\left| S - \prod_{p \leq x} \sum_{k=0}^{\infty} f(p^{-k}) \right| \leq \left| {\sum_{n \in A_x}}' f(n) \right| < \sum_{n > x} |f(n)| \to 0$$

as $x \to \infty$, whence $S = P$. Applying this result to $|f(n)|$, the absolute convergence of P follows, completing the proof.

We remark that the argument in Exercise 54 gives Euler's proof of the infinitude of primes. For if there are only finitely many primes, then (3.70) with $s = 1$ is over all primes, and (3.71) with $s = 1$ is over all positive integers:

$$\prod_{p \le x}(1 - p^{-1})^{-1} = \sum_{n=1}^{\infty} n^{-1} = \infty,$$

a contradiction. Needless to say, the first and the shortest proof of the infinitude of primes is due to Euclid, which may be stated in one line:

If $p_1, \cdots, p_n$ are all the primes, then $p_1 \cdots p_n + 1$ is also a prime.

Regarding the density of primes, much more is known. Letting, as usual

$$\pi(x) = \text{the number of primes } \le x,$$

the **prime counting function**, then the celebrated **prime number theorem** states that

$$\pi(x) \sim \frac{x}{\log x} \tag{3.72}$$

(cf. e.g. [Dave]).

Exercise 55. Use a multiplicative version of the Eratosthenes sieve (see (3.76) below) to prove (3.69)

Solution. We clearly have for $\sigma > 1$

$$(1 - 2^{-s})\zeta(s) = \sum_{2 \nmid n} \frac{1}{n^s}$$

the sum being extended over all positive odd integers. Similarly,

$$(1 - 2^{-s})(1 - 3^{-s})\zeta(s) = \sum \frac{1}{n^s},$$

the sum being extended over all integers not divisible by 2 or 3. Hence eventually,

$$\prod_{p}(1 - p^{-s})\zeta(s),$$

is extended over all integers not divisible by any prime p (the UFD property of $\mathbb{Z}$), whence it must be 1:

$$\prod_{p}(1 - p^{-s})\zeta(s) = 1,$$

which is (3.69).

The following "**inclusion-exclusion principle**" (cf. [Rio1, Chap. 3]) is sometimes the underlying principle of important identities: Let $A_1, \cdots, A_r$ be the subsets of a set A and $B = A \setminus \cup_{i=1}^r A_i$. Then

$$\sharp B = \sharp A + \sum_{s=1}^r (-1)^s \sum_{\{i_1, \cdots, i_s\} \subset \{1, \cdots, r\}} \sharp(A_{i_1} \cap \cdots \cap A_{i_s}). \qquad (3.73)$$

We show below that the **Eratosthenes sieve** is a manifestation of the inclusion-exclusion principle. Before that we state

Exercise 56. For distinct primes p, q, prove by the inclusion-exclusion principle

$$\varphi(pq) = pq - p - q + 1. \qquad (3.74)$$

Note that (3.74) is a basis for the RSA cryptology.

Solution. Let $X = \{1, \cdots, n\}$, $n = pq$ and let P resp. Q be the subset of X consisting of integers divisible by p resp. q. Then we want to find the number of elements in $X \setminus (P \cup Q)$. Since

$$\sharp(P \cup Q) = \sharp P + \sharp Q - \sharp(P \cap Q)$$

and $\sharp P = \left[\frac{n}{p}\right] = q$, $\sharp Q = \left[\frac{n}{q}\right] = p$, and $\sharp(P \cap Q) = 1$, it follows that

$$\sharp X \setminus (P \cup Q) = n - p - q + 1.$$

Theorem 3.13. (Eratosthenes sieve) *For $x > 0$ we have*

$$\pi(x) - \pi(\sqrt{x}) = [x] - 1 + \sum_{n | P} \mu(n) \left[\frac{x}{n}\right], \qquad (3.75)$$

where P is the product of all primes $\leq \sqrt{x}$.

This is a restatement of (3.73) in the form:

$$\pi(x) - \pi(\sqrt{x}) = [x] - 1 + \sum_{s=1}^{\pi(x)} (-1)^s \sum_{\substack{p_{i_1} \cdots p_{i_s} \leq x \\ \text{distinct}}} \left[\frac{x}{p_{i_1} \cdots p_{i_s}}\right]. \qquad (3.76)$$

For example, with $x = 100$ gives $\pi(100) = 25$.

3.6 The hyperbola method

In order to improve the error term for the Dirichlet divisor problem, we introduce the **hyperbola method** due to Dirichlet himself. This in fact is another manifestation of the inclusion-exclusion principle (3.73) as with Exercise 56.

The method will turn out to be very useful in deriving asymptotic formulas for the sum of other arithmetic functions, too.

Let $\rho = \rho(x)$, $0 < \rho \le x$ to be chosen suitably and consider the dissection of the sum

$$A(x) = \sum_{mn \le x} a(m)b(n)$$

$$= \sum_{m \le \rho} a(m) \sum_{n \le \frac{x}{m}} b(n) + \sum_{n \le \rho^{-1}x} b(n) \sum_{m \le \frac{x}{n}} a(m) - \sum_{m \le \rho, n \le \rho^{-1}x} a(m)b(n).$$

If $a(m) = b(n) = 1$, then

$$D(x) = \sum_{mn \le x} 1 = \sum_{n \le x} d(n)$$

$$= \sum_{m \le \rho} \sum_{n \le \frac{x}{m}} 1 + \sum_{n \le \rho^{-1}x} \sum_{m \le \frac{x}{n}} 1 - \sum_{m \le \rho} \sum_{n \le \rho^{-1}x} 1$$

$$= \sum_{m \le \rho} \left[\frac{x}{m}\right] + \sum_{n \le \rho^{-1}x} \left[\frac{x}{n}\right] - [\rho]\left[\rho^{-1}x\right].$$

Now use $[x] = x - \overline{B}_1(x) + \frac{1}{2}$ for $x \notin \mathbb{Z}$,

$$D(x) = \sum_{m \le \rho} \left(\frac{x}{m} - \overline{B}_1\left(\frac{x}{m}\right) + \frac{1}{2}\right) + \sum_{n \le \rho^{-1}x} \left(\frac{x}{n} - \overline{B}_1\left(\frac{x}{n}\right) + \frac{1}{2}\right)$$

$$- \left(\rho - \overline{B}_1(\rho) + \frac{1}{2}\right)\left(\frac{x}{\rho} - \overline{B}_1\left(\frac{x}{\rho}\right) + \frac{1}{2}\right)$$

$$= x \sum_{m \le \rho} \frac{1}{m} + x \sum_{n \le \rho^{-1}x} -x + \rho\left(\overline{B}_1\left(\frac{x}{\rho}\right) - \frac{1}{2}\right) - \frac{x}{\rho}\left(\overline{B}_1(\rho) - \frac{1}{2}\right)$$

$$- \sum_{m \le \rho} \overline{B}_1\left(\frac{x}{m}\right) - \sum_{n \le \rho^{-1}x} \overline{B}_1\left(\frac{x}{n}\right) + O(1),$$

if we use the first approximation

$$= x\left(\log\rho + \gamma + O\left(\frac{1}{\rho}\right)\right) + x\left(\log\frac{x}{\rho} + \gamma + O\left(\frac{\rho}{x}\right)\right) - x - \sum_{m\leq\rho}\overline{B}_1\left(\frac{x}{m}\right)$$

$$- \sum_{n\leq\rho^{-1}x}\overline{B}_1\left(\frac{x}{n}\right) + \rho\overline{B}_1\left(\frac{x}{\rho}\right) + \frac{x}{\rho}\overline{B}_1\left(\rho\right) + O(1)$$

$$= x\log x + (2\gamma - 1)x + O\left(\frac{x}{\rho}\right) + O\left(\frac{\rho}{x^2}\right) - \sum_{m\leq\rho}\overline{B}_1\left(\frac{x}{m}\right) - \sum_{n\leq\rho^{-1}x}\overline{B}_1\left(\frac{x}{n}\right)$$

$$+ \rho\overline{B}_1\left(\frac{x}{\rho}\right) + \frac{x}{\rho}.$$

To make $\frac{x}{\rho} = \rho$ hold, we are to chose $\rho = \sqrt{x}$.

$$D(x) = x\log x + (2\gamma - 1)x + O(\sqrt{x}),$$

i.e. Dirichlet's result. However, if we use the second approximation

$$\sum_{n\leq x}\frac{1}{n} = \log x + \gamma - B_1(x)\frac{1}{x} + O\left(\frac{1}{x^2}\right),$$

then,

$$D(x) = x\left(\log\rho + \gamma - B_1(\rho)\frac{1}{\rho} + O\left(\frac{1}{\rho^2}\right)\right) + x\left(\log\frac{x}{\rho} + \gamma - B_1(x)\frac{\rho}{x} + O\left(\frac{\rho^2}{x^2}\right)\right)$$

$$- x - \sum_{m\leq\rho}\overline{B}_1\left(\frac{x}{m}\right) - \sum_{n\leq\rho^{-1}x}\overline{B}_1\left(\frac{x}{n}\right) + \rho\overline{B}_1\left(\frac{x}{\rho}\right) + \frac{x}{\rho}\overline{B}_1\left(\rho\right) + O(1)$$

$$= x\log x + (2\gamma - 1)x + \Delta(x)$$

where

$$\Delta(x) = \sum_{m\leq\rho}\overline{B}_1\left(\frac{x}{m}\right) + \sum_{n\leq\rho^{-1}x}\overline{B}_1\left(\frac{x}{n}\right) + O\left(\frac{x}{\rho^2} + \frac{\rho^2}{x}\right)$$

$$= -2\sum_{n\leq\sqrt{x}}\overline{B}_1\left(\frac{x}{n}\right) + O(1)$$

for $\rho = \sqrt{x}$.

Using the Fourier series $-\frac{1}{\pi}\sum_{n=1}^{\infty}\frac{\sin 2\pi nx}{n}$ for $\overline{B}_1(x)$ and the estimate for trigonometrical sums, we can prove that

$$\Delta(x) = O(x^{\frac{1}{3}+\varepsilon})$$

or of smaller order.

3.7 Applications of Stieltjes integrals

We note that most of the results presented in §§3.2, 3.6 may be unified as those for summatory functions of Dirichlet convolutions (Definition 3.2). In this respect, it is known that one can use the notion of Stieltjes integrals to express the results in a lucid way. Cf. [Apo4] and [Widd] for a rather complete theory of Stieltjes integrals.

Definition 3.8. For bounded functions f, g defined on the interval $[a, b]$ one introduces the Stieltjes integral in almost verbatim to that of the Riemann integral. The only difference is that one uses g for the difference $x_{j+1} - x_j$: $g(x_{j+1}) - g(x_j)$,

$$\int_a^b f(x) \, dg(x).$$

Theorem 3.14. (i) *The Stieltjes integral exists if f is continuous and g is of bounded variation. The role can be changed in view of Item* (ii).
(ii) *The formula for integration by parts holds true.*

$$\int_a^b f(x) \, dg(x) = [f(x)g(x)]_a^b - \int_a^b g(x) \, df(x), \qquad (3.77)$$

provided that f is continuous and g is of bounded variation or g is continuous and f is of bounded variation.

(iii) *If g is a step function with jumps a_n at x_n, the Stieltjes integral reduces to the sum:*

$$\int_a^x f(x) \, dg(x) = \sum_{a < x_n \le x} f(x_n)a_n. \qquad (3.78)$$

(iv) *If f is continuous and g is differentiable, then the Stieltjes integral reduces to the Riemann integral:*

$$\int_a^b f(x) \, dg(x) = \int_a^b f(x)g'(x) \, dx. \qquad (3.79)$$

J. P. Tull ([Tul1]-[Tul4]) developed a general method for obtaining asymptotic formulas for the summatory function of the convolution of two arithmetic functions $a(n)$ and $b(n)$ whose summatory functions $A(x)$ and $B(x)$ satisfy asymptotic formulas. Indeed, his method is more general and can treat the Stieltjes resultant: Given two functions A and B defined for $x \ge 1$ of bounded variation on each bounded interval, one defines the **Stieltjes resultant** C of A and B by

$$C(x) = (A \times B)(x) = \int_a^x A(x/u) \, dB(u), \qquad (3.80)$$

whenever the integral exists and for all x, $C(x)$ lies between the limits $\lim_{h \to \mp 0} C(x \pm h)$. If $A(1) = B(1) = 0$, then (3.80) may be also written as $C(x) = (B \times A)(x) = \int_a^x B(x/u) \, \mathrm{d}A(u)$. Hence it is better to define the summatory function as

$$A(x) = \sum_{n < x} a(n).$$

Cf. Widder [Widd, pp. 83-91].

Example 3.5. We derive the special case of (3.33) for the Riemann zeta-function by the Stieltjes integral:

$$\zeta(s) = \frac{1}{s-1} + \frac{1}{2} - \int_1^\infty \overline{B_1}(x) x^{-s-l} \, \mathrm{d}x. \qquad (3.81)$$

In general suppose $f(x)$ is a continuous function on $[a, b]$. Then consider the integral

$$I = -\int_a^b \overline{B_1}(x) f(x) \, \mathrm{d}x.$$

By Theorem 3.14, (ii),

$$I = -\left[\overline{B_1}(x) f(x)\right]_a^b + \int_a^b f(x) \, \mathrm{d}\overline{B_1}(x) \qquad (3.82)$$

$$= -\left[\overline{B_1}(x) f(x)\right]_a^b + \int_a^b f(x) \, \mathrm{d}x - \int_a^b f(x) \, \mathrm{d}[x].$$

Since by Theorem 3.14, (iii),

$$\int_a^b f(x) \, \mathrm{d}[x] = \sum_{n - [a]+1}^{[b]} f(n), \qquad (3.83)$$

it follows that

$$\sum_{n - [a]+1}^{[b]} f(n) = \int_a^b f(x) \, \mathrm{d}x - \overline{B_1}(b) f(b) + \overline{B_1}(a) f(a) \qquad (3.84)$$

$$+ \int_a^b \overline{B_1}(x) \, \mathrm{d}f(x).$$

Now choose $a = 1$, $b = N \to \infty$ and $f(x) = x^{-s}$ with $\sigma > 1$. Noting that

$$\int_1^\infty x^{-s} \, \mathrm{d}x = \frac{1}{s-1},$$

we see that (3.84) leads to

$$\zeta(s) = \frac{1}{s-1} + \frac{1}{2} - \int_1^\infty \overline{B_1}(x)\,\mathrm{d}(x^{-s}),$$

which amounts to (3.81).

Since the integral in (3.81) is absolutely convergent for $\sigma > 0$, it defines an analytic function there, giving rise to an analytic continuation of $\zeta(s)$ to the wider half-plane $\sigma > 0$. On integrating, one gets an analytic continuation to $\sigma > -1$. Then the functional equation gives the meromorphic continuation to the whole complex plane, with the simple pole at $s = 1$ with residue 1.

Example 3.6. Example 3.5 gives the Laurent expansion

$$\zeta(s) = \frac{1}{s-1} + \sum_{n=0}^\infty \frac{\gamma_n}{n!}(s-1)^n, \tag{3.85}$$

where

$$\gamma_n = (-1)^n \lim_{N\to\infty} \left(\sum_{k=1}^N \frac{\log^n k}{k} - \frac{\log^{n+1} N}{n+1} \right) \tag{3.86}$$

is the n-th **generalized Euler constant** with γ_0 indicating the Euler constant γ ([Lehm]). We adopt Ferguson's proof [Ferg]. In view of (3.81), we have the Taylor expansion

$$H(s) := \zeta(s) - \frac{1}{s-1} + \sum_{n=0}^\infty A_n(s-1)^n \tag{3.87}$$

around $s = 1$. Since $n!A_n = H^{(n)}(1)$, we may differentiate (3.81) under the integral sign to get

$$n!A_n = \int_1^\infty \overline{B_1}(x)\,\mathrm{d}\left(\frac{\mathrm{d}^n}{\mathrm{d}s^n}\left(e^{-s\log x}\right) \right)\Bigg|_{s=1} = (-1)^{n+1} \int_1^\infty \overline{B_1}(x)\,\mathrm{d}\left(\frac{\log^n x}{x} \right).$$

Putting $a = 1$, $b = N \to \infty$ and $f(x) = x^{-s}$ in (3.84), we find that

$$n!A_n = (-1)^n \lim_{N\to\infty} \left(\sum_{k=1}^N \frac{\log^n k}{k} - \int_1^N \frac{\log^n x}{x}\,\mathrm{d}x \right), \tag{3.88}$$

which amounts to (3.86).

Cf. [Ivic] for the following general convolution result.

3.8   Characters as arithmetic functions

For basic knowledge on Dirichlet characters, cf. Chapter 1, §8. For a general theory of group characters, cf. Chapter 4.

We begin with

Example 3.7. ([Iwas, p. 4]) Let χ_1 and χ_2 be primitive characters with conductors f_1 and f_2, respectively. Then there is a unique primitive character with conductor f, $f \mid f_1 f_2$ such that

$$\chi(a) = \chi_1(a)\chi_2(a), \tag{3.89}$$

for $(a, f_1 f_2) = 1$. χ is called the product of χ_1 and χ_2 and denoted by: $\chi = \chi_1 \chi_2$. Note that (3.89) does not necessarily hold for $(a, f_1 f_2) > 1$.

Exercise 57. Prove that the set of all primitive characters forms a group under the multiplication (3.89).

Solution. The identity is given by the principal character χ_0^* mod 1 alluded to in Chapter 1, §8. The inverse is given by the complex conjugate $\bar{\chi}$: $\bar{\chi}(a) = \overline{\chi(a)}$.

The inclusion-"explosion" principle in §3.5 may be applied in various contexts. We dwell on the following. Let χ be a non-principal imprimitive Dirichlet character mod q induced by the primitive character χ^* with conductor f. We contend that

$$S_q := \sum_{m=1}^{q} \chi(m)m = \frac{q}{f} \prod_{\substack{p \mid q \\ p \nmid f}} (1 - \chi^*(p)) \sum_{m=1}^{f} \chi^*(m)m. \tag{3.90}$$

Indeed, since

$$S_q = \sum_{\substack{m=1 \\ (m,q)=1}}^{q} \chi(m)m$$

we deduce, by Remark 3.1, (i), that

$$S_q = \sum_{d \mid q} \mu(d)\chi^*(d)d \sum_{m'=1}^{q/d} \chi^*(m')m',$$

the sum now extending over $d \mid P$, where P is the product of all prime factors of q which do not divide f. Then d is of the form $p_1 \cdots p_k$ for which the inner sum of S_q is

$$\sum_{m'=1}^{q/d} \chi^*(m')m' = (-1)^k \sum_{p_1 \cdots p_k | q} \chi^*(p_1 \cdots p_k) p_1 \cdots p_k \sum_{m'=1}^{q/p_1 \cdots p_k} \chi^*(m)m$$

$$= (-1)^k \sum_{p_1 \cdots p_k | q} \chi^*(p_1 \cdots p_k) p_1 \cdots p_k \frac{q}{p_1 \cdots p_k f} \sum_{m=1}^{f} \chi^*(m)m.$$

Hence

$$S_q = \frac{q}{f} \sum_{m=1}^{f} \chi^*(m)m \left(\sum_{p_1 \cdots p_k | q} (-1)^k \chi^*(p_1 \cdots p_k) \right),$$

the inner sum of which is the product $\prod_{\substack{p|q \\ p \nmid f}} (1 - \chi^*(p))$, completing the proof of (3.90).

Exercise 58. Suppose χ is a non-principal imprimitive Dirichlet character mod q induced by the primitive character χ^* with conductor f. Let $L(s, \chi)$ be the associated Dirichlet L-function ((5.3), Chapter 5). Then prove that

$$L(s, \chi) = \prod_{p|q} \left(1 - \frac{\chi^*(p)}{p^s} \right) L(s, \chi^*), \tag{3.91}$$

where the product is over all prime divisors of q (which do not divide f).

Solution For $\sigma > 1$, $L(s, \chi) = \sum_{\substack{n=1 \\ (n,q)=1}}^{\infty} \frac{\chi^*(n)}{n^s}$, and so by Remark 3.1, (i)

$$L(s, \chi) = L(s, \chi^*) \sum_{d|q} \mu(d)\chi^*(d)d^{-s}, \tag{3.92}$$

the inner sum being $\sum_{d|P} \mu(d)\chi^*(d)d^{-s} = \prod_{p|P} \left(1 - \frac{\chi*(p)}{p^s} \right)$, where P is the same as above, completing the solution.

Chapter 4

Quadratic reciprocity through duality

In this chapter we shall develop duality theory which underlies number-theoretic setting but barely noticed by non-experts. It turns out that if we work with the dual space, we can discard a lot of superfluous arguments which are often presented elsewhere.

We shall give two proofs of the quadratic reciprocity. One depends on the Zolotareff-Frobenius theorem and the other on Hecke's reciprocity law in algebraic number fields as presented by L. Auslander, R. Tolimieri and S. Winograd [Ausl].

4.1 Group characters and duality

We begin with

Exercise 59. Let G be a group and let χ be a map $G \to \mathbb{C}$ having the multiplicativity:

$$\chi(xy) = \chi(x)\chi(y), \quad x, y \in G.$$

Prove that if χ is not the zero map 0, where the zero map is defined by

$$0(x) = 0, \quad \forall x \in G,$$

then

$$\chi(x) \neq 0, \quad \forall x \in G.$$

Solution. There is an $x \in G$ such that $\chi(x) \neq 0$. Hence

$$0 \neq \chi(x) = \chi(xe) = \chi(x)\chi(e),$$

whence

$$\chi(e) = 1. \tag{4.1}$$

Hence for any $x \in G$,
$$1 = \chi(e) = \chi(xx^{-1}) = \chi(x)\chi(x^{-1}),$$
whence the conclusion follows.

This also proves
$$\chi(x^{-1}) = \chi(x)^{-1} \qquad (4.2)$$
(cf. Proposition 1.4).

Hence except for the zero map, we may assume that χ is a homomorphism $\chi : G \to \mathbb{C}^*$, for which (4.1) and (4.2) hold.

If x is of finite order k, i.e. $x^k = e$, then
$$1 = \chi(x^k) = \chi(x)^k,$$
whence $\chi(x)$ is a k-th root of 1 and $|\chi(x)| = 1$.

Since we are mainly concerned with finite groups, all the elements are of finite order, and we may duly restrict the range of such homomorphisms to those with modulus 1.

Let $\mathbb{T} = \left\{ z \in G \middle| |z| = 1 \right\} = \left\{ e^{2\pi i\theta} \middle| \theta \in \mathbb{R} \right\}$ be the 1-dimensional torus group.

We call a homomorphism $\chi \in \text{Hom}(G, \mathbb{T})$ a **character** of G, and denote the set of all characters by $\hat{G}$.
$$\hat{G} = \text{Hom}(G, \mathbb{T}).$$

The map $\varepsilon(x) = 1$, for any $x \in G$ is in $\hat{G}$ and is called the **principal character**.

Theorem 4.1. *With the product $\chi\psi$ of $\chi, \psi \in \hat{G}$ defined by*
$$(\chi\psi)(x) = \chi(x)\psi(x), \ any \ x \in G, \qquad (4.3)$$
$\hat{G}$ becomes an Abelian group.

We refer to $\hat{G}$ as the **dual group** (or the **character group**) of G. Although our immediate application is to Dirichlet characters which are 0-extensions of Abelian characters of the multiplicative group of reduced residue classes modulo q, we shall develop a rather general theory of characters of (not necessarily finite) Abelian groups following somewhat [Galo], [Ausl] and others.

Proposition 4.1. *Let G be an Abelian group and let H be its proper subgroup. For $x \notin H$, let $H' = <H, x>$ denote the subgroup of G generated by H and x. Then a character χ on H can be extended to a character on H' with $\chi \neq \varepsilon$, where ε indicates the principal character.*

Further, if $(H' : H) = k$, then $\chi \in \hat{H}$ can be extended on H' in exactly k different ways.

Proof. If no powers of x belong to H save for $x^0 = e$, then we may define e.g. $\chi(x) = -1$ and

$$\chi(x^m h) = \chi(x)^m \chi(h) \quad for \ m \in \mathbb{Z}, \ h \in H \tag{4.4}$$

to get a character on H'. If a power of x, say x^k belongs to H, then we may choose the smallest positive integer k such that $x^k \in H$. This k is the index $(H' : H)$.

Since $\chi(x^k)$ is defined and is a k-th root of 1, we must choose the value of $\chi(x)$ to be one of k-th roots of 1.

Once the value of $\chi(x)$ is chosen, we may use (4.4) to extend χ on H'. In this way the choice of $\chi(x)$ gives k different characters. $\qquad\square$

Proposition 4.2. *Let G be an Abelian group and let H be a subgroup of G with a character χ. Then χ can be extended to a character on G.*

Proof. We apply Zorn's lemma. Let $\mathcal{A}$ be the set of all pairs (A, χ), where $A \supset H$ is a subgroup of G which has an extension of χ:

$$\mathcal{A} = \{(A, \chi) | H \subset a \subset G, \chi|_H = \chi\}.$$

Since $H \in \mathcal{A}$, we want to prove that $G \in \mathcal{A}$. We introduce the partial ordering $(A, \chi) \leqq (A', \chi')$ by the conditions A is a subgroup of A' and $\chi'|_A = \chi$.

Let $\mathcal{B}$ be a totally ordered subfamily of $\mathcal{A}$ and let

$$U = \cup_{A \in \mathcal{B}} A.$$

Then U is a subgroup of G that contains all $A \in \mathcal{B}$. Moreover, for any $x \in U$, there is a $(A, \chi) \in \mathcal{A}$ such that $x \in A$, so that $\chi(x)$ is defined. If there is another (A', χ'), then we may assume $(A, \chi) \leqq (A', \chi')$, and $\chi'(x) = \chi'|_A(x) = \chi(x)$, i.e. the value of $\chi(x)$ is independent of the choice of $(A, \chi) \in \mathcal{A}$. We may define a map χ on U in this way and show that it is multiplicative, and so $\chi \in \hat{U}$.

Hence (U, χ) is an upper bound of $\mathcal{B}$, whence $\mathcal{A}$ is inductive and Zorn's lemma applies.

Let (A, χ) be a maximal element of $\mathcal{A}$. Suppose there is an $x \notin A$. then by Proposition 4.1, $\chi \in \hat{A}$ can be extended to a character χ' on $A' = <A, x>$. Then $(A', \chi') \in \mathcal{A}$, and by $A \subsetneq A'$, we have $(A, \chi) < (A', \chi')$ a contradiction. $\qquad\square$

Proposition 4.1 also implies

Lemma 4.1. *If G is an Abelian group, then for any $x \neq 0$, there is a character χ such that $\chi \neq \varepsilon$.*

Exercise 60. Let G be a group and $\hat{G}$ be its character group. For a fixed element $x \in G$, define the new function

$$x : \hat{G} \to \mathbb{T}$$

by $x(\chi) = \chi(x)$, $\chi \in \hat{G}$. Prove that $x \in \hat{\hat{G}}$, which we denote by $\bar{x}$.

Proof. For $\chi, \psi \in \hat{G}$ we have

$$x(\chi\psi) = \chi\psi(x) = \chi(x)\psi(x) = x(\chi)x(\psi),$$

so that $x \in \mathrm{Hom}(\hat{G}, \mathbb{T})$, i.e. $x \in \hat{\hat{G}}$. $\qquad\qquad\square$

Exercise 61. Use Lemma 4.1 to show the map

$$\varphi : x \to \bar{x}, \ \bar{x} \in \hat{\hat{G}},$$

is an injective homomorphism.

Solution. For $x \in G$, $y \in G$ and $\chi \in \hat{G}$,

$$(\overline{xy})(\chi) = \chi(xy) = \chi(x)\chi(y)$$
$$= \bar{x}(\chi)\bar{y}(\chi) = (\bar{x}\bar{y})(\chi),$$

i.e. $\varphi(xy) = \varphi(x)\varphi(y)$. To prove that φ is injective, suppose $xy^{-1} \neq e$. Then by Lemma 4.1, there is a $\chi \in G$ such that $\chi(xy^{-1}) \neq 1$ or what amounts to the same thing, $(\overline{xy^{-1}})(\chi) \neq 1$, whence $\overline{xy^{-1}} \neq e$, i.e. $\bar{x} \neq \bar{y}$, whence φ is injective.

We are now in a position to prove the following theorem.

Theorem 4.2. *If G is a finite Abelian group, then $G \cong \hat{\hat{G}}$.*

Proof. By Exercise 61, we know that $G \cong \mathrm{Im}\,\varphi \subset \hat{\hat{G}}$ and so it suffices to prove that $|G| = |\hat{\hat{G}}|$. Since G is finite, we may start from $\{e\}$ to construct a sequence of subgroups H_i $(i = 0, 1, ..., r)$ such that

$$\{e\} = H_0 \subset H_1 \subset \cdots \subset H_{r-1} \subset H_r = G \qquad (4.5)$$

and $H_{i+1} = <H_i, x_i>$, with $x_i \notin H_i$. By Proposition 4.1, we may construct $(H_2 : H_1) \cdots (G : H_{r-1}) = |G|$ distinct characters on G and these exhaust all the elements of $\hat{G}$. It follows therefore that

$$|\hat{G}| = |G|. \qquad (4.6)$$

Applying this to $\hat{G}$, we obtain $|\hat{\hat{G}}| = |\hat{G}|$, which is $|G|$, we completing the proof. $\qquad\qquad\square$

Exercise 62. Prove the last lines leading to (4.6).

Solution. This is because each character χ is an extension of a character of some subgroup H of G, including G. Indeed, suppose χ is trivial on $G - H$, with H a subgroup and suppose $H_i \subset H \subset H_{i+1}$. Then χ must be the extension of χ_{H_i}. must be one of extensions of characters.

Exercise 63. Prove the **paring** $\chi(x) = (x, \chi) \in G \times \hat{G}$ is bilinear.

Solution. $(xy, \chi) = (x, \chi)(y, \chi)$ amounts to $\chi(xy) = \chi(x)\chi(y)$ (cf. the proof of Exercise 61), while $(x, \chi\psi) = (x, \chi)(y, \psi)$ amounts to $\chi\psi(x) = \chi(x)\psi(x)$ (cf. the proof of Exercise 60).

The following theorem is an improvement over Theorem 4.2 in the sense that $G \cong \hat{G}$ and $\hat{G} \cong \hat{\hat{G}}$ imply $G \cong \hat{\hat{G}}$.

Theorem 4.3. *If G is a finite Abelian group, then $G \cong \hat{G}$.*

The proof is divided into several steps.

Lemma 4.2. *For any pair of groups G_1, G_2 we have:*

$$\psi : \hat{G}_1 \times \hat{G}_2 \cong \widehat{G_1 \times G_2}; \quad \psi((\chi_1, \chi_2)) = \chi_1\chi_2,$$

where $\chi_1\chi_2 \in \widehat{G_1 \times G_2}$ is defined by

$$(\chi_1\chi_2)((x_1, x_2)) = \chi_1(x_1)\chi_2(x_2). \tag{4.7}$$

Proof. Since $(x_1, x_2)(x_1', x_2') = (x_1 x_1', x_2 x_2')$, we have

$$\begin{aligned}
(\chi_1\chi_2)((x_1, x_2)(x_1', x_2')) &= \chi_1(x_1 x_1')\chi_2(x_2 x_2') \\
&= \chi_1(x_1)\chi_2(x_2)\chi_1(x_1')\chi_2(x_2') \\
&= (\chi_1\chi_2)(x_1, x_2)(\chi_1\chi_2)(x_1', x_2'),
\end{aligned}$$

so that $\chi_1\chi_2 \in \widehat{G_1 \times G_2}$.

We show that any $\chi \in \widehat{G_1 \times G_2}$ may be written in the form $\chi_1\chi_2$. Indeed, defining $\chi_i \in \hat{G}_i$ by

$$\chi_1(x_1) = \chi((x_1, e_2)), \quad \chi_2(x_2) = \chi((e_1, x_2))$$

for any $x_i \in G_i$, where e_i is the identify element of G_i, we obtain

$$\begin{aligned}
\chi((x_1, x_2)) &= \chi((x_1, e_2))\chi((e_1, x_2)) = \chi_1(x_1)\chi_2(x_2) \\
&= (\chi_1\chi_2)((x_1, x_2)).
\end{aligned}$$

Hence ψ is onto.

If $\psi((\chi_1, \chi_2)) = \psi((\chi_1', \chi_2'))$, $\chi_i, \chi_i' \in \hat{G}_i$, then

$$\chi_1(x_1)\chi_2(x_2) = \chi_1'(x_1)\chi_2'(x_2)$$

for all $x_i \in G_i$. Choosing e.g. $x_1 = e_1$, we deduce $\chi_2 = \chi_2'$, and similarly, $\chi_1 = \chi_1'$. Hence ψ is one-to-one.

Finally, for any pairs $(\chi_1, \chi_2), (\chi_1', \chi_2') \in \hat{G}_1 \times \hat{G}_2$, we obtain

$$\psi((\chi_1, \chi_2))((\chi_1', \chi_2'))((x_1, x_2)) = \psi((\chi_1\chi_1', \chi_2\chi_2'))((x_1, x_2))$$
$$= (\chi_1\chi_1')(x_1)(\chi_2\chi_2')(x_2) = \chi_1(x_1)\chi_2(x_2)\chi_1'(x_1)\chi_2'(x_2)$$
$$= \psi((\chi_1, \chi_2))((x_1, x_2))\psi((\chi_1', \chi_2'))((x_1, x_2))$$
$$= (\psi((\chi_1, \chi_2)), \psi((\chi_1', \chi_2')))((x_1, x_2)).$$

Hence $\psi \in \mathrm{Hom}(\widehat{G_1 \times G_2}, \hat{G}_1 \times \hat{G}_2)$ and ψ is an isomorphism. $\square$

Exercise 64. If $G = <x>$ is a cyclic group of order n, then $G \cong \hat{G}$ under the isomorphism:

$$\psi(x^m) = \chi^m, \quad m = 1, \cdots, n,$$

where χ is defined by

$$\chi(x^k) = e^{2\pi i \frac{k}{n}}, \quad k = 1, \cdots, n. \tag{4.8}$$

Solution. The map χ defined by (4.8) is a character because

$$\chi(x^k x^h) = e^{2\pi i \frac{k+h}{n}} = \chi(x^k)\chi(x^h),$$

so are its powers $\chi^2, \cdots, \chi^n$. Since by (4.6), $|\hat{G}| = |G|$, it follows that ψ is an isomorphism.

The following lemma is a fundamental structure theorem for finite Abelian groups.

Lemma 4.3. *If G is a finite Abelian group, it is isomorphic to the direct product of cyclic groups of prime power order.*

This is a consequence of Theorem 2.5 with $R = \mathbb{Z}$. (Also in the proof of Lemma 10.1, Chapter 1, we obtained the decomposition into a direct product of groups of prime power order.)

Theorem 4.3 now follows from Exercise 64 and Lemmas 4.2 and 4.3.

In what follows we assume G is an Abelian group and the conclusion of Theorem 4.2 is true, i.e. $G \cong \hat{\hat{G}}$, although G is not assumed to be finite.

Exercise 65. Prove the equivalence:

$$\chi(x) = 1 \quad \text{for any} \quad \chi \in \hat{G}$$

if and only if $x = e$.

Solution. Recall the definition of the principal character ε:

$$\varepsilon(x) = 1 \quad \text{for any} \quad x \in G.$$

Hence $\chi(x) = 1$ for any $x \in G$ if and only if $\chi = \varepsilon$. Passing to the dual, we find that $x(\chi) = 1$ for any $\chi \in \hat{G}$ if and only if $x = \bar{e} = \psi(e)$, which we identify with e.

4.2 Finite Fourier transforms

Let G be a finite group of order g. We are concerned with the sum of the form

$$\frac{1}{g} \sum_{x \in G} \xi(x), \tag{4.9}$$

for any complex-valued function ξ on G. By interpreting $\frac{1}{g}$ to be the measure $\mu(x)$ of each element $g \in G$, we may think of (4.9) as an integral

$$\int_G \xi \, d\mu(x) = \int_G \xi,$$

and from now on we shall adopt the simple notation $\int_G \xi$:

$$\int_G \xi = \frac{1}{g} \sum_{x \in G} \xi(x). \tag{4.10}$$

Since G is finite, all complex-valued functions are bounded and integrable and form the group algebra $L^1(G)$. The multiplication is defined by the convolution

$$(\xi * \eta)(x) = \int_G \xi(x - y)\eta(y) \, d\mu(y). \tag{4.11}$$

In fact, G being finite, all other Banach spaces $L^2(G), L^\infty(G)$ are the same as $L^1(G)$. In $L^\infty(G)$, the multiplication is defined by pointwise multiplication, while $L^2(G)$, endowed with the **inner product**

$$(\xi, \eta) = \int_G \xi\bar{\eta} = \frac{1}{g} \sum_{x \in G} \xi(x)\bar{\eta}(x), \tag{4.12}$$

becomes a Hilbert space.

Let $a \in G$ be fixed. Then the left-translation $\xi_a \in L^1(G)$ of ξ is defined by

$$\xi_a(x) = \xi(ax). \tag{4.13}$$

Then we have

$$\int_G \xi_a = \xi, \quad a \in G, \tag{4.14}$$

i.e. all translations (including right translations) have the same average on G.

Indeed, (4.14) follows from the fact that along with x, ax also runs through all the elements of G.

Let $\chi \in \hat{G}$. Then its left-translation χ_a is given by $\chi_a = \chi(a)\chi$, since $\chi_a(x) = \chi(ax) = \chi(a)\chi(x) = (\chi(a)\chi)\,(x)$.

Hence, applying (4.14), we obtain

$$\chi(a) \int_G \chi(x) = \int_G \chi(x), \quad a \in G. \tag{4.15}$$

From (4.15) we deduce

Proposition 4.3.

$$\int_G \chi = \begin{cases} 0 & \chi \neq \varepsilon \\ 1 & \chi = \varepsilon \end{cases} \tag{4.16}$$

where ε is the principal character.

Proof.   Follows immediately by noting that the integral $\int_G \chi$ is 0 if and only if there is an element a of G such that $\chi(a) \neq 1$, i.e. if and only if $\chi \neq \epsilon$ or not. If $\chi = \epsilon$, then clearly, the integral is 1.   □

Proposition 4.3 has an important implication.

Corollary 4.1. (Orthogonality of characters) *For $\chi, \psi \in \hat{G}$,*

$$\sum_{x \in G} \chi(x)\bar{\psi}(x) = \begin{cases} 0 & \chi \neq \psi \\ 1 & \chi = \psi \end{cases} \tag{4.17}$$

and

$$\sum_{x \in \hat{G}} \chi(x)\bar{\chi}(y) = \begin{cases} 0 & x \neq y \\ 1 & x = y. \end{cases} \tag{4.18}$$

Proof.   If we replace χ by $\chi\bar{\psi}$ and noting that $\chi\bar{\psi} = \epsilon$ if and only if $\chi = \psi$, we obtain

$$\int_G \chi\bar{\psi} = \begin{cases} 0 & \chi \neq \psi \\ 1 & \chi = \psi, \end{cases} \tag{4.19}$$

which amounts to (4.17).

Working with the group $\hat{G}$, we may prove in a similar way to the above

$$\int_{\hat{G}} x\bar{y} = \begin{cases} 0 & x \neq y \\ 1 & x = y \end{cases} \tag{4.20}$$

where x, y are to be regarded as $\varphi(x), \varphi(y)$ in Exercise 3. (4.20) leads to (4.18). □

Note that if G is an Abelian group, then (4.20) is an exact duality relation of (4.19).

We define the **Fourier transform** $\hat{\xi}$ of $\xi \in L^2(G)$ by

$$\hat{\xi}(\chi) = \frac{1}{\sqrt{g}}(\xi, \chi) = \frac{1}{\sqrt{g}}\int_G \xi\bar{\chi}. \tag{4.21}$$

We now digress a little into the case of periodic arithmetic functions and prove a theorem of Hasse in a slightly generalized form.

For an arithmetical function $f(n)$ of period q, its (finite) Fourier transform (or discrete Fourier transform (DFT)) is defined by

$$\hat{f}(r) = \frac{1}{\sqrt{q}} \sum_{a=1}^{q} f(a)e^{-2\pi i \frac{r}{q}a}. \tag{4.22}$$

Then the **inversion formula**

$$\hat{\hat{f}}(r) = f(-r) \tag{4.23}$$

holds.

Exercise 66. Prove Formula (4.23).

Proof. The left-hand side is

$$\frac{1}{\sqrt{q}} \sum_{a=1}^{q} \hat{f}(a)e^{-2\pi i \frac{r}{q}a} = \frac{1}{q} \sum_{a=1}^{q} \sum_{b=1}^{q} f(b)e^{-2\pi i \frac{r}{q}a}e^{-2\pi i \frac{b}{q}a}$$

$$= \frac{1}{q} \sum_{b=1}^{q} f(b) \sum_{a=1}^{q} e^{-2\pi i \frac{r+b}{q}a}$$

$$= f(-r)$$

by the orthogonality of additive characters. □

The following theorem contains a generalization of the identity due to Hasse [Has1] (addendum by Newman; slightly generalized by Funakura [Funa, Theorem 10]). Its special case due to Dirichlet appears as Proposition 5.2, Chapter 2 and is used to determine the class number of the field in question.

Theorem 4.4. *Suppose q is an odd integer and let χ be a Dirichlet character modulo q. Then*

$$(1 - 2\chi(2)) \sum_{a=1}^{q} \chi(a)a = \chi(2)q \sum_{a=1}^{(q-1)/2} \chi(a). \tag{4.24}$$

Equation (4.24) *still remains true in the case q being even meaning that both sides are 0.*

For this we need some lemmas from Funakura [Funa].

Lemma 4.4.

$$\sum_{r=1}^{q-1} \hat{f}(r)(2r - q) = i\sqrt{q} \sum_{r=1}^{q-1} f(r) \cot \frac{r}{q}\pi. \tag{4.25}$$

Lemma 4.5.

$$\sum_{r=1}^{q-1} \hat{f}(r)(2r - q) = i\sqrt{q} \sum_{r=1}^{q-1} f(r) \cot \frac{r}{q}\pi. \tag{4.26}$$

Proof. The left-hand side becomes

$$\frac{1}{\sqrt{q}} \sum_{m=1}^{q} f(m) \sum_{n=1}^{q-1} \left((-1)^n e^{-\frac{2\pi imn}{q}} + (-1)^{n-q} e^{-\frac{2\pi im(n-q)}{q}} (n - q) \right) \tag{4.27}$$

$$= \frac{1}{\sqrt{q}} \sum_{m=1}^{q} f(m) \sum_{n=1}^{q-1} n \left(r^n - r^{-n} \right),$$

where $r = -e^{-\frac{2\pi im}{q}} = e^{2\pi i \left(-\frac{m}{q} - \frac{1}{2}\right)}$.

Hence $r^{\frac{1}{2}} = e^{\pi i \left(-\frac{m}{q} - \frac{1}{2}\right)}$ and $r^{-\frac{1}{2}} = e^{\pi i \left(\frac{m}{q} + \frac{1}{2}\right)}$.

By

$$\sum_{n=1}^{q-1} n \left(r^n - r^{-n} \right) = q \frac{r^{\frac{1}{2}} + r^{-\frac{1}{2}}}{r^{\frac{1}{2}} - r^{-\frac{1}{2}}},$$

the inner sum of (4.27) is

$$q \frac{-ie^{-\frac{\pi m}{q}i} + ie^{\frac{\pi m}{q}i}}{-ie^{-\frac{\pi m}{q}i} - ie^{\frac{\pi m}{q}i}} = q \frac{-2i \sin \frac{\pi m}{q}\pi}{2i \cos \frac{m}{q}\pi} = -qi \tan \frac{m}{q}\pi. \qquad \square$$

Suppose f is periodic of period of q, q odd, $f(q) = 0$
  f is even: $f(n - q) = f(n)$. Then

$$\sum_{\substack{n=1 \\ 2\nmid n}}^{q-1} f(n) = \sum_{\substack{n=1 \\ 2\nmid n}}^{q-1} f(q - n) = \sum_{\substack{m=1 \\ 2\mid m}}^{q-1} f(m)$$

because q is odd.
  Hence

$$\sum_{n=1}^{q-1} f(n) = \sum_{2\mid n} + \sum_{2\nmid n} = 2 \sum_{2\mid n} f(n).$$

Putting

$$S = \sum_{n=1}^{q-1} f(n),$$

we obtain

$$2S = \sum_{n=1}^{2(q-1)} f(n) = 2 \sum_{\substack{n=1 \\ 2\mid n}}^{2(q-1)} f(n) = 2 \sum_{n=1}^{q-1} f(2n).$$

Now let $f(n) = \chi \cot \frac{n}{q}$ and $2 \cot 2\pi i x = \cot \pi x - \tan \pi x$, we have

$$2S = 2 \sum_{n=1}^{q-1} \chi(2n) \cot \frac{2n}{q} \pi = \chi(2)S - \chi(2) \sum_{n=1}^{q-1} \chi(n) \tan \frac{n}{q} \pi.$$

Hence

$$(2(\bar{\chi})(2) - 1)S = - \sum_{n=1}^{q-1} \chi(n) \tan \frac{n}{q} \pi.$$

Lemma 4.6. (Parity Lemma) *For any function f, we have*

$$\sum_{n=1}^{q} (-1)^n f(n) = 2 \sum_{n=1}^{[\frac{q}{2}]} f(2n) - \sum_{n=1}^{q} f(n). \tag{4.28}$$

Proof. Proof follows immediately on classifying the values of a mod 2.$\Box$

Proof of Theorem 4.4. We recall [Vist, Lemma 8.1]: If f is odd, then $\sum_{n=1}^{q-1} f(a) = 0$ and if f is even, then $2 \sum_{a \leq q/2} f(a) = \sum_{a=1}^{q-1} f(a)$.
  Hence, if χ is even, then both sides of (4.24) are 0.
  Now suppose χ is odd. Then we find that

$$\sum_{n=1}^{q-1} \chi(n)(-1)^n = 2\chi(2) \sum_{n=1}^{\frac{q-1}{2}} \chi(n),$$

a consequence of Lemma 4.6.

Lemma 4.5 with $f = \hat{\chi}$ reads

$$\sum_{n=1}^{q-1} \chi\left((-1)^n n + (-1)^{n-q}(n-q)\right) = i\sqrt{q} \sum_{n=1}^{q-1} \hat{\chi}(n) \tan\frac{n}{q}\pi,$$

whose left-hand side is $-q\sum_{n=1}^{q-1} \chi(n)(-1)^n$. The right-hand side is $-(2\chi(2)-1)i\sqrt{q}\sum_{n=1}^{q-1}\hat{\chi}(n)\cot\frac{n}{q}\pi$, which is by Lemma 4.4,

$$-(2\chi(2)-1)\sum_{n=1}^{q-1}\chi(n)(2n-q) = -2(2\chi(2)-1)\sum_{n=1}^{q-1}\chi(n)n.$$

Comparing these, we complete the proof.

4.3   Quadratic reciprocity through Dedekind sums

In this section, we partly follow Rademacher-Grosswald [RaGr] to deduce the quadratic reciprocity law ((5.61) in Example 1.9; Theorem 5.6) from the **reciprocity law for Dedekind sums** and their relations to the Jacobi symbol furnished by the Zolotareff-Frobenius theorem. We begin by starting the definition of Dedekind sums. Here we are concerned with the most fundamental one $s(h, k)$ defined for $(h, k) = 1$, $1 < k \in \mathbb{Z}$ by

$$s(h, k) = \sum_{j=1}^{k}((j/k))((hj/k)), \qquad (4.29)$$

where $((x))$ is the special notation for Dedekind sums, meaning the periodic Bernoulli polynomial of degree 1 with the intermediate value at discontinuities.

Theorem 4.5. *Let h, k be relatively prime positive integers > 1. Then the reciprocity law holds true.*

$$s(h, k) + s(k, h) = \frac{1}{12}\left(\frac{h}{k} + \frac{k}{h} - \frac{1}{hk}\right) + \frac{1}{4}. \qquad (4.30)$$

Dedekind sums show interesting features under the action of the modular group (Exercise 32). Cf. [RaGr, Chapter 4, p. 45]. On [RaGr, p. 32], Dedekind's formula is proved

$$12s(h, k) \equiv k + 1 - 2\left(\frac{h}{k}\right) \pmod{8}, \qquad (4.31)$$

whence the quadratic reciprocity law is deduced on account of (4.30).

Another derivation of the quadratic reciprocity is to appeal to the relation [RaGr, p. 37] between the Dedekind sum and the symbol $Z(h, k)$ introduced in the following Theorem 4.6:

$$Z(h, k) = -3s(h, k) + \frac{1}{2}(k - 1)(k - 2), \qquad (4.32)$$

whence by (4.31), Theorem 4.6 follows for k odd.

Here we give a proof due to R. E. Dressler and E. E. Shult [DrSh].

Theorem 4.6. (Zolotareff-Frobenius theorem) *We define* $Z(h, k)$ *to be* sgn σ, *where* $\sigma \in S(\mathbb{Z}/k\mathbb{Z})$ *is a permutation on* $\mathbb{Z}/k\mathbb{Z}$ *obtained by multiplying by* h *and suppose* k *is odd. Then*

$$Z(h, k) = \left(\frac{h}{k}\right), \qquad (4.33)$$

the Jacobi symbol.

Let $k = p_1^{e_1} \cdots p_n^{e_n}$. Then (1.35) reads

$$\mathbb{Z}/k\mathbb{Z} \cong \mathbb{Z}/p_1^{e_1}\mathbb{Z} \times \cdots \times \mathbb{Z}/p_n^{e_n}\mathbb{Z}. \qquad (4.34)$$

Choose in Exercise 27 $X_i = \mathbb{Z}/p_i^{e_i}\mathbb{Z}$, $y_i = \frac{k}{p_i^{e_i}}$ and let σ_i be obtained by multiplying by h. Then (1.47) reads

$$Z(h, k) = \prod_{i=1}^{n} \operatorname{sgn} \sigma_i^{y_i} = \prod_{i=1}^{n} \operatorname{sgn} \sigma_i = Z(h, p_i^{e_i}), \qquad (4.35)$$

the second equality because y_i being odd.

This reduces the problem to the case of an odd prime power $k = p^e$. In this case, we apply the fact that the unit group $(\mathbb{Z}/p^e\mathbb{Z})^{\times}$ is cyclic.

Now let $X = \mathbb{Z}/p^e\mathbb{Z}$ and let $D_i = \{\bar{x} \in X | (\bar{x}, k) = p^i\}$, where $(\bar{x}, k)$ means the gcd (y, k) for $y \in \bar{x}$. Note that $\sharp D_i = (p-1)p^{e-1-i}$, $0 \le i \le e-1$.

Lemma 4.7. (Salié) *For* h, k *both odd positive, we have*

$$hZ(h, k) + kZ(k, h) = \frac{1}{4}(h - 1)(k - 1)(h + k - 1). \qquad (4.36)$$

This immediately follows from (4.36) and (4.30).

Proof of Theorem 5.6. Follows by combining Theorem 4.6 and Lemma 4.6.

4.4 Quadratic reciprocity in algebraic number fields

In this section we follow the presentation in [Ausl] and derive the quadratic reciprocity law in the same way as one can reach Hecke's theorem on the quadratic reciprocity law in algebraic number fields [Heck].

Let k be an algebraic number field of degree n and let $\mathrm{Tr} = Tr_{k/\mathbb{Q}}$ be the trace from k to $\mathbb{Q}$. k has r_1 real conjugate fields $k^{(j)}$, $1 \le j \le r_1$ and $2r_2$ imaginary conjugate fields $k^{(j)}, \overline{k^{(j)}}$, $1 \le i \le r_2$. For $\beta \in k^\times$ and $1 \le i \le r_1$, we define

$$\mathrm{sgn}_j(\beta) = \begin{cases} 1, & \beta^{(j)} > 0 \\ -1, & \beta^{(j)} < 0 \end{cases} \tag{4.37}$$

and

$$S(\beta) = \sum_{j=1}^{r_1} \mathrm{sgn}_j(\beta). \tag{4.38}$$

Let M be a full module in Definition 2.5, Chapter 2, i.e. $M = \mathbb{Z}\alpha_1 \oplus \cdots \oplus \mathbb{Z}\alpha_n$ with an integral basis $\{\alpha_i\}$ of k with respect to M. Let M' denote the **dual** of M with respect to the trace

$$M' = \{x \in k \,|\, \mathrm{Tr}(xM) \subset \mathbb{Z}\}. \tag{4.39}$$

By Lemma 2.1, there exists a $\mathbb{Q}$-basis $\{\alpha'_i\}$ of k dual to $\{\alpha_i\}$: $\mathrm{Tr}(\alpha_i \alpha'_j) = \delta_{ij}$. Hence M' is a full module with integral basis $\{\alpha'_i\}$.

Now choose any $\alpha \in k^\times$ and form $(\alpha M)'$. Then $(\alpha M)' = (1/\alpha)M'$. Indeed, if $\beta \in (\alpha M)'$, then $\mathrm{Tr}((\beta\alpha)M)\mathrm{Tr}(\beta(\alpha M)) \subset \mathbb{Z}$, whence $\beta\alpha \in M'$ and $\beta \in (1/\alpha)M'$. The reverse inequality follows by reversing the argument, proving the assertion.

Now let M^2 denote the set of all the products mn with $m, n \in M$. Then M^2 is a finitely generated $\mathbb{Z}$-module containing the full submodule mM where $m \ne 0$ is any fixed element of M, whence M^2 is a full module. Choose any $0 \ne \alpha \in (M^2)'$ and form $(\alpha M)'$. Then for any $\beta \in M$, we have $\mathrm{Tr}(\beta(\alpha M)) \subset \mathrm{Tr}((\alpha M^2)) \subset \mathbb{Z}$, whence $M \subset (\alpha M)'$. Hence we may form the quotient group, which is a finite Abelian group

$$A(\alpha) = A(\alpha, M) = (\alpha M)'/M. \tag{4.40}$$

Now we apply the above situation to the ring $M = \mathcal{O}$ of integers in k. It is a full module and $\mathcal{O}^2 = \mathcal{O}$, with $1/\mathcal{O}'$ the **different** of k. Hence for each $0 \ne \alpha \in \mathcal{O}$, we may form a finite Abelian group

$$A(\alpha) = A(\alpha, \mathcal{O}) = (\alpha\mathcal{O})'/\mathcal{O}. \tag{4.41}$$

For $\beta \in (\alpha M)'$ let

$$\chi(\beta)(\gamma) = e^{2\pi i \mathrm{Tr}(\alpha\beta\gamma)}, \quad \gamma \in (\alpha M)'.$$

Then $\chi(\beta) \in \mathrm{Hom}((\alpha M)', \mathbb{T})$ and $\chi(\beta)$ acts trivially on M. Hence $\chi(\beta)$ induces a character $D(\beta) \in \widehat{A(\alpha)}$. Then $D \in \mathrm{Hom}((\alpha M)', \widehat{A(\alpha)})$ satisfies the short exact sequence

$$0 \to M \to (\alpha M)' \xrightarrow{D} \widehat{A(\alpha)} \to 1, \tag{4.42}$$

which simply means that

$$(\alpha M)'/\mathrm{Ker}\, D \simeq \widehat{A(\alpha)} \tag{4.43}$$

and $\mathrm{Ker}\, D = M$, the latter of which we now prove. Indeed, $\beta \in \mathrm{Ker}\, D$ means that $D(\beta)(c) = 1$ for $\forall c \in A(\alpha)$ and thence that $e^{\mathrm{Tr}(\alpha\beta\gamma)} = \chi(\beta)(\gamma) = 1$ for $\forall \gamma \in (\alpha M)'$, which implies that $\mathrm{Tr}(\alpha\beta(\alpha M)') \subset \mathbb{Z}$. Since $(\alpha M)' = (1/\alpha)M'$, it follows that $\mathrm{Tr}(\beta M') \subset \mathbb{Z}$, i.e. $\beta \in M'' = M$. $\mathrm{Ker}\, D \supset M$ being clear, we conclude the identity. Since the order of $A(\alpha) = \widehat{A(\alpha)}$, we conclude (4.43).

The short exact sequence (4.42) implies that D induces an isomorphism (4.43), which we denote by $D(\alpha)$. It is determined by

$$(D(\alpha)(c))(b) = (b, D(\alpha)(c)) = e^{2\pi i \mathrm{Tr}(\alpha b c)}, \quad b, c \in A(\alpha), \tag{4.44}$$

where $(a, \hat{a}) \in A(\alpha) \times \widehat{A(\alpha)}$ is a bilinear paring in Exercise 63.

Exercise 67. Prove that the last equality of (4.44) is well-defined.

Solution. If $b = \beta + M$, $c = \gamma + M$, $b, c \in (\alpha M)'$, then

$$e^{2\pi i \mathrm{Tr}(\alpha b c)} = e^{2\pi i \mathrm{Tr}(\alpha\beta\gamma)}$$

in view of (4.39).

For the moment we denote $A(\alpha)$ by A. Let δ_a denote the character defined by

$$\delta_a(b) = \begin{cases} 1 & b = a \\ 0 & b \neq a \end{cases}. \tag{4.45}$$

Then $\{\delta_a | a \in A\}$ forms an orthonormal basis of $L^2(A)$ with respect to the inner product

$$(f, g) = \sum_{a \in A} f(a)\bar{g}(a). \tag{2.4'}$$

Since $(f, \delta_a) = f(a)$, we have the Fourier expansion

$$f = \sum_{a \in A} (f, \delta_a)\delta_a = \sum_{a \in A} f(a)\delta_a. \qquad (4.46)$$

Recall that the canonical isomorphism in Theorem 4.2 is given by φ
$(\varphi(a)(\chi) = \chi(a))$ in Exercise 61. From Corollary 4.1 it follows that

$$\left\{ \frac{1}{\sqrt{m}}\varphi(x) \Big| x \in A \right\} \qquad (4.47)$$

is an orthonormal basis of $L^2(\hat{A})$.

We introduce the Fourier transform $\mathcal{F} = \mathcal{F}_A : L^2(A) \to L^2(\hat{A})$ as the unique linear operator satisfying

$$\mathcal{F}(\delta_a) = \frac{1}{\sqrt{m}}\varphi(a). \qquad (4.48)$$

Then its concrete form is given by (cf. (4.22))

$$\mathcal{F}(f)(\chi) = \frac{1}{\sqrt{m}} \sum_{x \in A} (x, \chi)f(x), \quad \chi \in \hat{A}, f \in L^2(A). \qquad (4.49)$$

Indeed, substituting (4.46), we see that $\mathcal{F}(f) = \sum_{x \in A} f(x)\mathcal{F}(\delta_x)$, which leads to (4.49) in view of (4.48).

Let $D^\sharp$ denote the linear isometry given by $D^\sharp(\chi) = \chi \cdot D^\sharp, \quad \chi \in L^2(\hat{A})$. We consider the unitary operator $D^\sharp \cdot \mathcal{F}(f)$ of $L^2(A)$, which satisfies

$$D^\sharp \cdot \mathcal{F}(\delta_b) = \frac{1}{\sqrt{m}} \sum_{c \in A} (D(b))(c)\delta_c, \quad b \in A. \qquad (4.50)$$

Exercise 68. Prove (4.50).

Solution. We are to check the values of both sides at $b \in A$. By definition,

$$D^\sharp \cdot \mathcal{F}(\delta_a) = \mathcal{F}(\delta_a)(D(b)) = \frac{1}{\sqrt{m}} \sum_{c \in A} (c, D(b))\delta_a(c) \qquad (4.51)$$

by (4.49). The far-right member is $\frac{1}{\sqrt{m}}(a, D(b))$ by the definition of δ_a, which is $\frac{1}{\sqrt{m}}(b, D(a))$ by symmetry. By (4.46), we have

$$D(a) = \sum_{c \in A} D(a)(c)\delta_c.$$

Substituting this, we find that

$$D^\sharp \cdot \mathcal{F}(\delta_a)(b) = \frac{1}{\sqrt{m}}(b, D(a))$$

$$= \frac{1}{\sqrt{m}} \sum_{c \in A} D(a)(c)(b, \delta_c) = \frac{1}{\sqrt{m}} \sum_{c \in A} D(a)(c)\delta_c(b),$$

which is (4.50).

The unitary operator (4.50) determines the "Fourier transform" $D^\sharp \cdot \mathcal{F}$ on the basis $\{\delta_a\}$ of A and the corresponding matrix is given by

$$\frac{1}{\sqrt{m}}(D(a))(c)) = \frac{1}{\sqrt{m}}((c, D(a)))_{a,c \in A} \tag{4.52}$$

and its trace is the **Gauss sum**

$$G(\alpha) = G(A, \alpha) = \frac{1}{\sqrt{m}} \sum_{a \in A} (a, D(a)). \tag{4.53}$$

In view of (4.44), we have $(c, D(b)) = e^{2\pi i \mathrm{Tr}(abc)}$, whence (4.50) and (4.53) in the case $A = A(\alpha)$ the finite Abelian group defined in (4.41) read respectively

$$F(\alpha) = \frac{1}{\sqrt{m(\alpha)}} \sum_{c \in A(\alpha)} e^{2\pi i \mathrm{Tr}(abc)} \delta_c, \quad b \in A(\alpha), \tag{4.54}$$

$$G(\alpha) = G(A(\alpha), \alpha) = \frac{1}{\sqrt{m(\alpha)}} \sum_{b \in A(\alpha)} e^{2\pi i \mathrm{Tr}(ab^2)} \tag{4.55}$$

where $F(\alpha) = D^\sharp \cdot \mathcal{F} = D^\sharp(\alpha) \cdot \mathcal{F}_{A(\alpha)}$, $G(\alpha) = G(A, \alpha)$ and $m(\alpha)$ is the order of $A(\alpha)$.

In [Ausl], there are two assumptions made which are eventually proved, but here we take them for granted.

Milgram formula:

$$G(4\alpha) = 2^n e^{2\pi i S(\alpha)/8} \tag{4.56}$$

and

$$\mathcal{O}' = \delta \cdot \mathcal{O} \tag{4.57}$$

for some $\delta \in k$.

Milgram's formula is essential in the proof of the Hecke reciprocity law, and in [Ausl], it is proved by studying finite Fourier transforms as intertwining operators between two representations of a nilpotent group. These intertwining operators are described by the multi-dimensional Cooley-Tukey algorithm for computing the finite Fourier transform.

Let $\mathfrak{p}$ be a prime ideal in $\mathcal{O}$. Then according to Remark 2.1, $\mathcal{O}/\mathfrak{p}$ is a finite field, say F. Hence the theory of finite fields (§1.10, Chapter 1) applies and we introduce the Legendre symbol with respect to $\mathfrak{p}$ in the same way: For $a \in F^\times$, let

$$\left(\frac{a}{\mathfrak{p}}\right) = \begin{cases} 1 & \text{if } a \text{ is a square in } F^\times \\ -1 & \text{otherwise} \end{cases}, \tag{4.58}$$

which is a character of $F^\times$. We extend the definition to $\alpha \in \mathcal{O}, \notin \mathfrak{p}$ by setting $(a = \alpha + \mathfrak{p} \in F^\times)$

$$\left(\frac{\alpha}{\mathfrak{p}}\right) = \left(\frac{a}{\mathfrak{p}}\right). \tag{4.59}$$

Let $0 \neq \alpha \in \mathcal{O}'$. Then since $(\alpha\mathcal{O})' = (1/\alpha)\mathcal{O}'$, it follows that $a := \alpha/\delta \in \mathcal{O}$ in view of (4.57). Hence any member of $\mathcal{O}$ may be written in this form.

Now fix $0 \neq \alpha \in \mathcal{O}'$ and take $\beta/\delta \in \mathcal{O}$ relatively prime to α/δ in $\mathcal{O}$: $(\alpha/\delta)\mathcal{O} + (\beta/\delta)\mathcal{O} = \mathcal{O}$. The mapping $\gamma \to (\beta/\delta)\gamma$ is seen to be an isomorphism of $(\alpha\mathcal{O})'$, which induces an automorphism $\zeta_\alpha(\beta)$. For $b \in (\mathcal{O}/a\mathcal{O})^\times$, we have a representation $b = \eta/\delta + (\alpha/\delta)\mathcal{O}$ by which we may define $\zeta_\alpha(b) := \zeta_\alpha(\beta)$.

Lemma 4.8. *Suppose $b \in (\mathcal{O}/a\mathcal{O})^\times$, where $a = \alpha/\delta$. Then*

$$G(\alpha; b) := \frac{1}{\sqrt{m(\alpha)}} \sum_{c \in A(\alpha)} e^{2\pi i \mathrm{Tr}(\alpha b c^2)}, \tag{4.60}$$

which is the trace of the regular representation $F(\alpha)\zeta_\alpha(b) = \zeta_\alpha(b)^{-1} F(\alpha)$.

The last equality ([Ausl, Lemma 3.2]) gives ([Ausl, Cor. 3.2.4])

$$G(\alpha; bc) = G(\alpha; c) \tag{4.61}$$

if $b, c \in (\mathcal{O}/(\alpha/\delta)\mathcal{O})^\times$ and b is a square. It also holds that

$$G(\alpha; b) = G(\alpha) \tag{4.62}$$

if b is a square in $(\mathcal{O}/(\alpha/\delta)\mathcal{O})^\times$.

The trigonometric sum in (4.60) is referred to as a **generalized Hecke sum**. Let $a = \alpha/\delta$ be an odd prime (i.e. a is relatively prime to 2) in $\mathcal{O}$ and write $\left(\frac{b}{a}\right) = \left(\frac{b}{a\mathcal{O}}\right)$ for $b \notin a\mathcal{O}$.

Theorem 4.7.

$$\left(\frac{b}{a}\right) = \frac{G(\alpha; b)}{G(\alpha)}. \tag{4.63}$$

Proof. Let g be a generator of $(\mathcal{O}/(\alpha/\delta)\mathcal{O})^{\times} = (\mathcal{O}/a\mathcal{O})^{\times}$ of order m. Then

$$\left(\frac{g^j}{a}\right) = (-1)^j.$$

By (4.61), $G(\alpha; g^j)$ is $G(\alpha)$ or $G(\alpha; g)$ according as j is even or odd. Hence it suffices to prove that $G(\alpha; g^j) = -G(\alpha)$. Since

$$A(\alpha) = (\alpha\mathcal{O})'/\mathcal{O} = (1/\alpha)(\mathcal{O})'/\mathcal{O} = (1/\alpha)(1/\delta\mathcal{O})/\mathcal{O} = (1/a)(\mathcal{O})/\mathcal{O} \cong \mathcal{O}/a\mathcal{O},$$

(4.55) and (4.60) read

$$\sqrt{m(\alpha)}G(\alpha) = 1 + \sum_{j=0}^{m-1} e^{2\pi i \mathrm{Tr}((\delta/a)g^{2j})}$$

and

$$\sqrt{m(\alpha)}G(\alpha; g) = 1 + \sum_{j=0}^{m-1} e^{2\pi i \mathrm{Tr}((\delta/a)g^{2j+1})},$$

respectively.

Adding these, we see that $\sqrt{m(\alpha)}(G(\alpha) + G(\alpha; g))$ is $\sum_{b \in \mathcal{O}/a\mathcal{O}} e^{2\pi i \mathrm{Tr}((\delta/a)b)}$, which is 0 by the orthogonality relation Proposition 4.3, $e^{2\pi i \mathrm{Tr}((\delta/a)b)}$ being a non-trivial character of $\mathcal{O}/a\mathcal{O}$, completing the proof. $\qquad\square$

Lemma 4.9. ([Ausl, Cor. 3.51]) *If α/δ and β/δ are relatively prime in $\mathcal{O}$, then*

$$G\left(\frac{\alpha\beta}{\delta}\right) = G(\alpha; b)G(\beta; a), \tag{4.64}$$

where $a = \alpha/\delta + (\beta/\delta)\mathcal{O} \in (\mathcal{O}/(\beta/\delta)\mathcal{O})^{\times}$ and b is similar.

Lemma 4.10. ([Ausl, Cor. 3.6.1]) *If α/δ is odd and a square mod $4\mathcal{O}$, then $G(4\delta : \alpha/\delta) = G(4\delta)$ and*

$$G(\alpha) = \frac{G(4\alpha)}{G(4\delta)} = e^{\frac{\pi}{4}i(S(\alpha)-S(\delta))}. \tag{4.65}$$

Theorem 4.8. *Assume (4.57). Then for non-associated odd primes $a, b \in \mathcal{O}$, we have*

$$\left(\frac{b}{a}\right)\left(\frac{a}{b}\right) = e^{\frac{\pi i}{4}\mathrm{Tr}(\delta(ab-a-b))}\frac{G(4;a)G(4;b)}{G(4;ab)}. \tag{4.66}$$

Further, if a is a square mod $4\mathcal{O}$, then

$$\left(\frac{b}{a}\right)\left(\frac{a}{b}\right) = e^{\pi i \sum_{j=1}^{r_1} \frac{\mathrm{sgn}_j(a)-1}{2}\frac{\mathrm{sgn}_j(b)-1}{2}}. \tag{4.67}$$

Proof. By Theorem 4.7, we obtain

$$\left(\frac{b}{a}\right)\left(\frac{a}{b}\right) = \frac{G(b\delta; a)}{G(b\delta)}\frac{G(a\delta; b)}{G(a\delta)} = \frac{G(ab\delta)}{G(b\delta)G(a\delta)} \tag{4.68}$$

by Lemma 4.9. Hence by Lemma 4.10, it follows that

$$\left(\frac{b}{a}\right)\left(\frac{a}{b}\right) = \frac{G(4ab\delta)}{G(4b\delta)G(4a\delta)}\frac{G(4\delta; b)G(4\delta; a)}{G(4\delta; ab)}. \tag{4.69}$$

By the Milgram formula (4.56), the first factor on the right-hand side of (4.69) gives the first factor of the right-hand side of (4.66).

Equation (4.67) follows from Lemma 4.10, completing the proof. □

Remark 4.1. We have given only a sketchy proof of the simplest case here. The interested reader should consult the paper [Ausl] for the general case. This section is in good contrast to §2, Chapter 5, where the theta-transformation formula is used, especially interesting and essential is the difference in the evaluation of the quadratic Gauss sums. Although not cited in [Ausl], there is another significant paper [TKub] which deals with clarification of Hecke's argument from the point of view of unitary representations, which can be read along with the paper [Ausl] with benefit. It is noted that the author claims he does not use the functional equation (cf. [TKub, p. 48]). For an extensive account of reciprocity laws, cf. [Lemm].

Chapter 5

Around Dirichlet L-functions

Since most of the essential facts about Dirichlet L-functions take the simplest form for those with primitive characters, we devote the first half of this chapter to presenting basic facts in that case, the general case of imprimitive characters being briefly sketched toward the end of the chapter.

We shall also give intriguing results which more or less follow from the functional equation for the Dirichlet L-function. The first is a derivation of the quadratic reciprocity law from the theta transformation formula, which in turn is equivalent to the functional equation for the Riemann zeta-function. The second is the Szmidt-Urbanowicz-Zagier result on short interval character sums with polynomial weight and their relation to class numbers of the associated quadratic fields. It is remarkable that their result depends on the associated Lambert series, which is again equivalent to the functional equation.

We also state Yamamoto's result which treats the case of Clausen function weight as well as the Bernoulli polynomial weight.

Finally, we state a mean square formula for the Dirichlet L-function at negative integer points.

5.1 Dirichlet L-functions with primitive characters

Let

$$G_n(\chi) = \sum_{a=1}^{q} \chi(a) e^{2\pi i \frac{n}{q} a} \tag{5.1}$$

denote the (generalized) **Gauss sum** and let $\tau(\chi)$ be the normalized Gauss sum:

$$\tau(\chi) = G_1(\chi) = \sum_{a=1}^{q} \chi(a) e^{2\pi i \frac{a}{q}}. \tag{5.2}$$

For an extensive theory of these Gauss sums, cf. §5.4 as well as below.

The **Dirichlet *L*-function** (sometimes *L*-series) $L(s, \chi)$ is defined by

$$L(s, \chi) = \sum_{n=1}^{\infty} \frac{\chi(n)}{n^s}, \quad \sigma = \operatorname{Re} s > 1. \tag{5.3}$$

As most of the zeta- and *L*-functions, the Dirichlet *L*-function satisfies the functional equation, which takes the following concise form for χ a *primitive* character:

$$L(1 - s, \chi) = q^{s-1}(2\pi)^{-s}\Gamma(s) \left(e^{-\frac{\pi i}{2}s} + \chi(-1)e^{\frac{\pi i}{2}s} \right) \tau(\chi)L(s, \bar{\chi}), \tag{5.4}$$

which is usually stated in the form (see e.g. [Dave, Chapter 9])

$$\xi(1 - s, \chi) = \frac{i^{\mathfrak{a}}\sqrt{q}}{\tau(\chi)}\xi(s, \bar{\chi}), \tag{5.5}$$

where

$$\xi(s, \chi) = \pi^{-\frac{s+\mathfrak{a}}{2}}\Gamma\left(\frac{s + \mathfrak{a}}{2}\right) L(s, \chi), \tag{5.6}$$

and where $\mathfrak{a}$ is 0 or 1 according as $\chi(-1) = 1$ or $\chi(-1) = -1$ ($\chi(-1) = (-1)^{\mathfrak{a}}$):

$$\mathfrak{a} = \mathfrak{a}(\chi) = \frac{1 - \chi(-1)}{2} = \begin{cases} 0 & \chi \text{ even} \\ 1 & \chi \text{ odd} \end{cases}. \tag{5.7}$$

Exercise 69. Prove that the two expressions (5.4) and (5.5) are equivalent.

Solution. Use is made of two properties of the gamma function.

Two most famous independent proofs are known for the functional equation, one depending on the theta transformation formula (cf. Exercise 71 below) and the other on the Hurwitz formula through the very convenient expression

$$L(s, \chi) = q^{-s} \sum_{a=1}^{q-i} \chi(a)\zeta\left(s, \frac{a}{q}\right), \tag{5.8}$$

where $\zeta(s, \alpha)$ indicates the Hurwitz zeta-function defined for $0 < \alpha \le 1$ by

$$\zeta(s, \alpha) = \sum_{n=1}^{\infty} \frac{1}{(n + \alpha)^s}, \quad \sigma > 1, \tag{5.9}$$

whose special case is the Riemann zeta-function

$$\zeta(s) = \sum_{n=1}^{\infty} \frac{1}{n^s}, \quad \sigma > 1. \tag{5.10}$$

Both of them are analytic in the right half-plane in the first instance and can be meromorphically continued to the whole complex plane. The functional equation for the Riemann zeta-function $\zeta(s)$ (as a consequence of (5.5)) with the even character χ_0^* reads

$$\pi^{-\frac{s}{2}}\Gamma\left(\frac{s}{2}\right)\zeta(s) = \pi^{-\frac{1-s}{2}}\Gamma\left(\frac{1-s}{2}\right)\zeta(1-s). \tag{5.11}$$

For an asymmetric form, cf. (5.74) below.

We shall give a variant of the theta-transformation proof through the Lipschitz summation formula below, which is of interest in its own right. Indeed, on the way, we also achieve the proof depending on the Hurwitz zeta-function. For $\sigma > 1$, (5.8) is stated as [Vist, (8.17), p. 171]. Note that for $0 < \sigma \leq 1$, the series (5.3) for $L(s,\chi)$ (χ non-trivial) is uniformly convergent in s, so that it is analytic in $\sigma > 0$. In the remaining range, the functional equation (cf. (5.12) below) for the Dirichlet L-function gives its analytic continuation and (5.8) is valid for all $s \neq 1$.

In contrast to (5.4), the formula

$$L(1-s,\chi) = q^{s-1}(2\pi)^{-s}\Gamma(s)\left(e^{-\frac{\pi i}{2}s} + \chi(-1)e^{\frac{\pi i}{2}s}\right)\ell(s,\bar{\chi}) \tag{5.12}$$

is valid for *all* χ, where

$$\ell(s,\chi) = \sum_{n=1}^{\infty} G_n(\chi)n^{-s} = \sum_{a=1}^{q-1}\chi(a)\ell_s\left(\frac{a}{q}\right) \tag{5.13}$$

is the ℓ-function considered by many authors including Joris [Jori], Neukirch [Neuk], Kubert-Lang [KuLa] et al. Here $\ell_s(x) = \sum_{n=1}^{\infty}\frac{e^{2\pi ixn}}{n^s}$ is the **polylogarithm function**, a special case $L(x,s,1)$ of the **Lipschitz-Lerch transcendent** which we introduce by (([SrCh, (11), p. 122]))

$$L(s,x,a) = \sum_{n=0}^{\infty}\frac{e^{2\pi ixn}}{(n+a)^s} \quad \sigma > 0,\ \alpha \in \mathbb{C}-\{0,-1,-2,\cdots\},\ \mu \in \mathbb{R}\backslash\mathbb{Z}. \tag{5.14}$$

Then the Hurwitz formula (the functional equation for the Hurwitz zeta-function; see Corollary 5.1 below) is a consequence of the functional equation for the Lipschitz-Lerch transcendent stated in Theorem 5.3 below:

$$L(1-s,x,a) = \frac{\Gamma(s)}{(2\pi)^s}\left\{e^{\pi i(\frac{1}{2}s-2ax)}L(s,-a,x)\right. \tag{5.15}$$
$$\left. +e^{-\pi i(\frac{1}{2}s-2a(1-x))}L(s,a,1-x)\right\}$$

valid for $0 < x < 1$ (([SrCh, (10), p. 122])).

In order to mention the known special values of the zeta- and L-functions at integer arguments, we appeal to ordinary Bernoulli numbers (defined by (3.36)). We mention the relation between them and Leopoldt's Bernoulli numbers $\tilde{B}_n$ ([HLeo]), which are defined by

$$\sum_{n=0}^{\infty} \tilde{B}_n \frac{t^n}{n!} = \frac{te^t}{e^t - 1} = \frac{t}{e^t - 1} + t = \sum_{n=0}^{\infty} B_n \frac{t^n}{n!} + t, \quad |t| < 2\pi, \qquad (5.16)$$

whence

$$\tilde{B}_1 = \frac{1}{2} = -B_1, \ \tilde{B}_k = B_k \ (k \neq 1). \qquad (5.17)$$

Suppose χ is a primitive character with conductor $f = f_\chi$. Note that in Iwasawa [Iwas] and [Wash], all the characters are assumed to be primitive (cf. [Iwas, p.4, ll.1-2], [Wash, Was]). Then **Leopoldt's generalized Bernoulli numbers** $B_{n,\chi}$ are introduced through the generating function

$$F_\chi(t) = \sum_{a=1}^{f} \frac{\chi(a)te^{at}}{e^{ft} - 1} = \sum_{n=0}^{\infty} B_{n,\chi} \frac{t^n}{n!}, \quad |t| < 2\pi/f. \qquad (5.18)$$

[Wash, Proposition 4.1, p. 31] gives a closed form for $B_{n,\chi}$:

$$B_{n,\chi} = F^{n-1} \sum_{a=1}^{F} \chi(a) B_n \left(\frac{a}{F}\right), \qquad (5.19)$$

for any multiple F of f, where $B_n(x)$ indicates the n-th Bernoulli polynomial

$$B_n(x) = (B + x)^n = \sum_{k=0}^{n} \binom{n}{k} B_k x^{n-k}, \qquad (5.20)$$

and where the middle term of (5.20) obeys umbral calculus.

Leopoldt's generalized Bernoulli polynomial $B_{n,\chi}(x)$ may be introduced by

$$B_{n,\chi}(x) = \sum_{k=0}^{n} \binom{n}{k} B_{k,\chi} x^{n-k} \qquad (5.21)$$

which is the special case $r = 1$ of $B_{n,\chi}^{[r]}(x)$ defined by (5.83).

Since $B_1(x) = x - \frac{1}{2}$, we have for non-principal χ, $B_{1,\chi} = \frac{1}{f} \sum_{a=1}^{f} \chi(a)a$, but $B_{1,\chi_0^*} = \frac{1}{2}$. Clearly, $B_{0,\chi} = \frac{\varphi(q)}{q}$ if χ is the principal character, $B_{0,\chi} = 0$ otherwise.

Theorem 5.1. (Leopoldt) *For any positive integer n and (primitive) character χ, we have the evaluation*

$$L(1 - n, \chi) = -\frac{B_{n,\chi}}{n}, \qquad (5.22)$$

whence

$$L(n, \chi) = (-1)^{1 + \frac{n-\mathfrak{a}}{2}} \frac{\tau(\chi)}{2i^{\mathfrak{a}}} \left(\frac{2\pi}{f} \right)^n \frac{B_{n,\bar{\chi}}}{n!}, \qquad (5.23)$$

for $n \equiv \mathfrak{a} \mod 2$.

In particular, for $n \in \mathbb{N}$, $\zeta(1-n) = -\frac{B_n(1)}{n}$ or $\zeta(1-n) = -\frac{B_n}{n}$ for $n > 1$ and $\zeta(0) = -\frac{1}{2}$ (cf. (5.122) below) and for even $n \geq 2$, we have Euler's formula ([Eule]: cf. Proposition 5.1 below)

$$\zeta(n) = (-1)^{1 + \frac{n}{2}} \frac{1}{2} (2\pi)^n \frac{B_n}{n!}, \qquad (5.24)$$

whence follows e.g. $\zeta(2) = \frac{\pi^2}{6}$, Euler's solution to the Basler problem.

Thus which notation we choose for Bernoulli numbers is rather immaterial as far as closed expressions for special values are concerned (save for the very special case of χ_0^*).

In what follows we shall deduce (5.15) from the Lipschitz summation formula, which in turn is a consequence of the Poisson summation formula. First we recall the latter as stated in Rademacher [Rade] (for a character analogue cf. (5.41) below).

Lemma 5.1. (Poisson summation formula) *Suppose $f \in C^2(\mathbb{R})$ and that integrals*

$$\int_{-\infty}^{\infty} f(x) \, dx \quad and \quad \int_{-\infty}^{\infty} |f''(x)| \, dx \qquad (5.25)$$

exist. Let

$$a_n = \hat{f}(n) = \int_{-\infty}^{\infty} f(x) e^{-2\pi i n x} dx \qquad (5.26)$$

*be the Fourier transform of f. Then we have the **Poisson summation formula***

$$S(u) := \sum_{n=-\infty}^{\infty} f(n+u) = \sum_{n=-\infty}^{\infty} a_n e^{2\pi i n u} \qquad (5.27)$$

uniformly in u in any finite interval, and in particular

$$S(0) := \sum_{n=-\infty}^{\infty} f(n) = \sum_{n=-\infty}^{\infty} \hat{f}(n). \qquad (5.28)$$

Theorem 5.2. (Lipschitz summation formula)*For the complex variables* $z = x + iy$, $x > 0$, $s = \sigma + it$, $\sigma > 1$ *and the real parameter* $0 < \alpha \leq 1$, *we have the Lipschitz summation formula*

$$\frac{(2\pi)^s}{\Gamma(s)} \sum_{m=0}^{\infty} (m+\alpha)^{s-1} e^{-2\pi z(m+\alpha)} = \sum_{n=-\infty}^{\infty} \frac{e^{2\pi in\alpha}}{(z+ni)^s}. \qquad (5.29)$$

Under the condition $0 < \alpha < 1$ *this formula holds in the wider half-plane* $\sigma > 0$.

Proof. The left-hand side series is absolutely convergent for all s in view of the convergence factor $e^{-2\pi x(m+\alpha)}$.

For $\sigma > 1$, the right-hand series is also absolutely convergent because the n-th term of a Majorant series is $O\left(\frac{1}{|(n+1)^\sigma|}\right)$.

The Poisson summation formula applied to $f(u) = \dfrac{e^{2\pi iu\alpha}}{(z+ui)^s}$ gives

$$S(0) = \sum_{n=-\infty}^{\infty} f(n) = \sum_{m=-\infty}^{\infty} A_m,$$

where

$$A_m = \int_{-\infty}^{\infty} \frac{e^{2\pi iu\alpha}}{(z+ui)^s} e^{-2\pi imu}\, du.$$

Putting $z + ui = w$, we obtain

$$A_m = \frac{1}{i} \int_{z-i\infty}^{z+i\infty} \frac{e^{2\pi(w-z)(\alpha-m)}}{w^s}\, dw = \frac{e^{2\pi z(m-\alpha)}}{i} \int_{z-i\infty}^{z+i\infty} \frac{e^{-2\pi w(m-\alpha)}}{w^s}\, dw,$$

or more conveniently

$$A_{-m} = \frac{e^{-2\pi z(m+\alpha)}}{i} I_m,$$

say, where

$$\int_{z-i\infty}^{z+i\infty} \frac{e^{2\pi w(m+\alpha)}}{w^s}\, dw. \qquad (5.30)$$

First we treat the terms with $m + \alpha > 0$. Putting $2\pi w(m+\alpha) = \zeta$, we obtain

$$A_{-m} = \frac{e^{-2\pi z(m+\alpha)}(2\pi)^s (m+\alpha)^{s-1}}{2\pi i} \int_{a-i\infty}^{a+i\infty} \frac{e^\zeta}{\zeta^s}\, d\zeta,$$

where we put $a = 2\pi(m+\alpha)\,\mathrm{Re}(z)$. By the **Laplace expression** for the gamma function

$$\frac{1}{\Gamma(s)} = \int_{(c)} \frac{e^z}{z^s}\, dz, \qquad (5.31)$$

where (c) means the vertical line $z = c + iy$, $-\infty < y < \infty$ with $c > 0$, to be proved in Exercise 70 below, we have

$$A_{-m} = \frac{e^{-2\pi z(m+\alpha)}(2\pi)^s(m+\alpha)^{s-1}}{\Gamma(s)}.$$

In the remaining case $m + \alpha \leq 0$, the integrals (5.30) are 0.
Thus we deduce Formula (5.29).   □

Exercise 70. For $\sigma > 0$, prove the Laplace expression (5.31) for the gamma function.

Solution. Recall the definition of the gamma function as the Mellin transform.

$$\Gamma(s) = \int_0^\infty e^{-t} t^{s-1} \, dt$$

for $\sigma > 0$. For elements of Mellin and Laplace transforms, cf. Chapter 6 and [Vist, Chapter 7].

Make the change of variable $t = zu$ with $\operatorname{Re} z > 0$ to obtain

$$\frac{\Gamma(s)}{z^s} = \int_0^\infty e^{-zt} t^{s-1} \, dt. \tag{5.32}$$

More precisely, the procedure is as follows. First we assume $z > 0$ and then we check that both sides are analytic in $\operatorname{Re} z > 0$ and so they must coincide. Indeed, the left-hand side is analytic in the cut plane $-\pi < \arg z \leq \pi$.

Now we think of (5.32) as the Laplace transform of the function t^{s-1}. Then by the inversion formula

$$t^{s-1} = \frac{1}{2\pi i} \int_{(c)} e^{zt} \frac{\Gamma(s)}{z^s} \, dz, \quad t > 0$$

or

$$\frac{1}{\Gamma(s)} = \frac{1}{2\pi i} \int_{(c)} e^{zt} \frac{1}{(zt)^s} \, d(zt),$$

which is (5.31).

Another solution is possible by applying the Hankel expression.

Theorem 5.2 as it stands refers to the summation of the divergent series $\sum_{m=0}^\infty (m+\alpha)^{s-1}$ by means of the convergence factor $e^{-2\pi z(m+\alpha)}$, $x > 0$.

It leads, however, to a more essential proposition which is so crucial in the whole spectrum of number theory:

Theorem 5.3. *For* $0 < \alpha < 1$, $0 < \mu < 1$ *we have the Lerch functional equation* ((5.15))

$$e^{2\pi i \mu \alpha} L(1 - s, \mu, \alpha) \tag{5.33}$$

$$= \frac{\Gamma(s)}{(2\pi)^s} \left(e^{\frac{\pi i}{2} s} L(s, 1 - \alpha, \mu) + e^{-\frac{\pi i}{2} s + 2\pi i \alpha} L(s, 1 - \alpha, \mu) \right).$$

Remark 5.1. Apostol [Apo1] has determined the values of $L(x, s, a)$ at non-positive integral arguments:

$$L(x, -k, a) = -\frac{\mathcal{B}_{k+1}(a, e^{2\pi i x})}{k + 1}, \quad k = 0, 1, \cdots, \tag{5.34}$$

where $\mathcal{B}_n(a, \alpha)$ are **Apostol's generalized Bernoulli polynomials** defined by the generating function

$$\frac{z e^{az}}{\alpha e^z - 1} = \sum_{n=0}^{\infty} \mathcal{B}_n(a, \alpha) \frac{z^n}{n!}, \quad |z + \log \alpha| < 2\pi \tag{5.35}$$

$(B_n(a) = \mathcal{B}_n(a, 1))$ is the n-th Bernoulli polynomial; cf. [Apo1] and [SrCh, pp. 126-127]).

In [SrCh, p. 126], Apostol (1951b) should be read Apostol (1951b) [Apo2], since there is another paper by him (1951a) [Apo1]. In Apostol (1952) [Apo3], it is stated that L. Carlitz pointed out the source of the generalized Bernoulli polynomial $\mathcal{B}(a.\alpha)$.

Exercise 71. Deduce (5.33) from (5.29).

Solution. The right-hand side of (5.29) may be written for $\sigma > 1$ as

$$e^{\frac{\pi i}{2} s} \sum_{n=0}^{\infty} \frac{e^{2\pi i (1 - \alpha) n}}{(n + \mu)^s} + e^{-\frac{\pi i}{2} s} \sum_{n=0}^{\infty} \frac{e^{2\pi i \alpha (n+1)}}{(n + 1 - \mu)^s},$$

which is the second factor of the right-hand side of (5.33).

On the other hand, the left-hand side of (5.29) is

$$\frac{(2\pi)^s}{\Gamma(s)} e^{2\pi i \mu \alpha} L(1 - s, \mu, \alpha)$$

for $\sigma < 1$.

Now by Dirichlet's test, the series (5.14) is uniformly convergent for $\sigma > 0$ and represents an analytic function there. It follows that Formula

(5.33) valid in the common domain $0 < \sigma < 1$ must hold true for all s but for singularities of $\Gamma(s)$. This completes the proof of (5.33).

Corollary 5.1. *For $0 < \alpha < 1$ the Hurwitz formula holds:*

$$\zeta(1 - s, \alpha) = \frac{\Gamma(s)}{(2\pi)^s} \left(e^{\frac{\pi i}{2} s} \ell_s(1 - \alpha) + e^{-\frac{\pi i}{2} s} \ell_s(\alpha) \right). \tag{5.36}$$

We are now in a position to deduce the functional equation (5.4) from Corollary 5.1.

Proof of (5.4). Substituting (5.36) in (5.8), we deduce that

$$L(1-s,\chi) = q^{s-1} \frac{\Gamma(s)}{(2\pi)^s} \left(e^{\frac{\pi i}{2} s} \sum_{a=1}^{q-1} \chi(a) \ell_s \left(1 - \frac{a}{q} \right) + e^{-\frac{\pi i}{2} s} \sum_{a=1}^{q-1} \chi(a) \ell_s \left(\frac{a}{q} \right) \right),$$

which amounts to (5.4), completing the proof.

Theorem 5.4. (Apostol) *The separability of the Gauss sum*

$$G_n(\chi) = \bar{\chi}(n)\tau(\chi) \tag{5.37}$$

characterizes the primitivity of χ.

Proof. If $(n,q) = 1$, then $\chi(a) = \chi(an)\bar{\chi}(n)$ and so

$$G_n(\chi) = \bar{\chi}(n) \sum_{a=1}^{q} \chi(an) e^{2\pi i \frac{na}{q}} = \bar{\chi}(n)\tau(\chi) \tag{5.38}$$

since an runs through all residue classes mod q. If now $(n,q) = 1$ and χ is primitive, then we can show that both sides of (5.38) are 0. Hence (5.37) holds for χ primitive.

Conversely, assume (5.37). Then by Lemma 5.6, $\tau(\chi) = G_1(\chi) \neq 0$ if and only $R = 1$, i.e. if and only if $\frac{q}{f}$ is square-free and $\left(\frac{q}{f}, f \right) = 1$, and then

$$\tau(\chi) = \mu \left(\frac{q}{f} \right) \psi \left(\frac{q}{f} \right) \tau(\psi) \tag{5.39}$$

and

$$G_n(\chi) = g(n)\tau(\chi), \tag{5.40}$$

where $g(n)$ is given by (5.91).

Now under (5.37), for $G_n(\chi) \neq 0$, we must have $\tau(\chi) \neq 0$, which is the case if and only if $R = 1$ and $g(n) = \bar{\chi}(n)$ for $\forall n \in \mathbb{N}$, i.e. $\tilde{q} = 1$, or $q = f$, i.e. χ is primitive. $\qquad\square$

We note that (5.39) is true for all characters. Indeed, if $\frac{q}{f}$ is not square-free or $\left(\frac{q}{f}, f\right) > 1$, then the right-hand side of (5.39) is 0 and so $\tau(\chi) = 0$.

Exercise 72. Prove that if the Dirichlet L-function $L(s, \chi)$ satisfies the functional equation (5.4), then χ is primitive.

First solution. [Ber1, Theorem 2.3] states that if f is of bounded variation on $[a, b]$, $-\infty < a < b < \infty$, then the **character analogue of the Poisson summation formula**

$$\frac{1}{2} \sideset{}{'}\sum_{a \leq n \leq x} \chi(n)(f(n+0) + f(n-0)) \tag{5.41}$$

$$= \frac{4\tau(\chi)}{q} \sum_{n=1}^{\infty} \int_a^b f(u)\bar{\chi}(n) \frac{e^{2\pi i \frac{u}{q} n} + \chi(-1)e^{-2\pi i \frac{u}{q} n}}{2i^{\mathfrak{a}}} \, du$$

holds, where $\mathfrak{a}$ is defined in (5.7).

If we choose $f(u) = e^{2\pi i \frac{m}{q} n}$ in (5.41), then we obtain (5.4), whence by Apostol's theorem, χ is primitive.

Second solution. (Joris) Formula (5.12) is true for all χ with $\ell(s, \chi)$ being the second member in (5.13). Comparing (5.4) and (5.12) yields (5.37) again.

Third solution. Here we appeal to the universal formula (5.116). If $\tilde{q} > 1$, then by (5.117), $L(s, \chi)$ has a non-real zero on the imaginary axis, which is a zero of $L(s, \overline{\chi})$ on $\sigma = 1$, contrary to the zero-free result. Hence $\tilde{q} = 1$ and $L(s, \chi) = L(s, \psi)$. Hence because of (5.4), we must have $q = f$, and χ is primitive.

5.2 The quadratic reciprocity

In this section we follow the lines of [MuPa] to deduce the Gauss quadratic reciprocity from the theta transformation formula, which is in the long run equivalent to the functional equation for the associated zeta-function (i.e. the Riemann zeta-function).

Theorem 5.5. *Suppose that $f(x, t)$ is integrable in x on $[a, b]$ for all t near the point α and that the limit function $\lim_{t \to \alpha} f(x, t) = f(x, \alpha)$ is integrable on $[a, b]$. Further suppose that the limit is uniform in $x \in [a, b]$. Then we may take the limit under the integral sign:*

$$\lim_{t \to \alpha} \int_a^b f(x, t) \, dx = \int_a^b \lim_{t \to \alpha} f(x, t) \, dx. \tag{5.42}$$

Proof. By the uniformity of convergence, given $\varepsilon > 0$, there exists a $\delta = \delta(\varepsilon) > 0$ such that for $0 < |t - \alpha| < \delta$,

$$|f(x,t) - f(x,\alpha)| < \varepsilon \tag{5.43}$$

for all $x \in [a, b]$. Hence

$$\left| \int_a^b f(x,t)\,\mathrm{d}x - \int_a^b f(x,\alpha)\,\mathrm{d}x \right| \leq \varepsilon \int_a^b \mathrm{d}x = (b-a)\varepsilon,$$

which completes the proof. $\qquad\square$

Corollary 5.2. *Suppose that $f(x,t)$ is integrable in x on $[a,b]$ for all t in a certain domain T, that $f(x,t)$ is differentiable in $t \in T$ for every $x \in [a,b]$ and that $f_t = \frac{\partial f}{\partial t}$ is continuous in both the variables x and t. Then we may differentiate under the integral sign:*

$$\frac{\mathrm{d}}{\mathrm{d}t} \int_a^b f(x,t)\,\mathrm{d}x = \int_a^b \frac{\partial}{\partial t} f(x,t)\,\mathrm{d}x. \tag{5.44}$$

Proof. By Theorem 5.5, it suffices to prove that

$$\lim_{h \to 0} \frac{\Delta f}{h} = f_t(x,t)$$

uniformly in $x \in [a,b]$, where $\Delta f = f(x,t+h) - f(x,t)$. By the mean value theorem, there exists a θ, $0 \leq \theta \leq 1$ such that

$$\Delta f = f(x,t+h) - f(x,t) = f_t(x,t+\theta h).$$

Since $f_t(x,t)$ is continuous in $x \in [a,b]$, it is uniformly continuous on $[a.b]$. Hence if $0 < |h|$ is small enough, then

$$\left| \frac{\Delta f}{h} - f_t(x,t) \right| = |f_t(x,t+\theta h) - f_t(x,t)|$$

is arbitrarily small uniformly in $x \in [a,b]$. Hence the convergence is uniform, and the proof is complete. $\qquad\square$

Example 5.1. (Theta transformation formula) Let $\theta(t)$ denote the Jacobi elliptic theta function

$$\theta(t) = \sum_{n=-\infty}^{\infty} e^{-\pi n^2 t}, \quad t > 0. \tag{5.45}$$

Then it satisfies the transformation formula

$$\theta\left(\frac{1}{t}\right) = t^{1/2}\theta(t). \tag{5.46}$$

Proof.   In Lemma 5.1 we choose $f(x) = e^{-\pi n^2 x}$ and evaluate the Fourier coefficients.

$$\hat{f}(n) = \int_{-\infty}^{\infty} e^{-\pi x^2 t} e^{-2\pi i n x} \mathrm{d}x. \tag{5.47}$$

The integrand may be expressed as $e^{-\pi n^2/t} e^{-\pi(x\sqrt{t}+in/\sqrt{t})^2}$.   Hence it amounts to establishing

$$\int_{-\infty}^{\infty} e^{-\pi(x+iu)^2} \mathrm{d}x = 1 \tag{5.48}$$

for any $u \in \mathbb{R}$, which is done in Exercise 73 below.

Under (5.48), we have $\hat{f}(n) = t^{1/2} e^{-\pi n^2/t}$. Hence (5.28) leads to (5.46), completing the proof.   □

Exercise 73. Prove (5.48).

Solution. We assume the value of the probability integral (cf. e.g. [Vist, pp. 32, 34])

$$\int_{-\infty}^{\infty} e^{-\pi x^2} \mathrm{d}x = 1 \tag{5.49}$$

as known and show that the value of the integral in (5.48) is independent of the value of u.

Differentiating the integral in (5.48) under the integral sign, which we may because of absolute convergence, we see that the result is

$$\int_{-\infty}^{\infty} 2\pi i(x + iu) e^{-\pi(x+iu)^2} \mathrm{d}x, \tag{5.50}$$

which we may view as

$$i \int_{-\infty}^{\infty} \frac{\partial}{\partial x} e^{-\pi(x+iu)^2} \mathrm{d}x. \tag{5.51}$$

Since this is simply $\left[ie^{-\pi(x+iu)^2}\right]_{-\infty}^{\infty}$, which is 0, and we have shown that the integral in (5.48) is independent of the value of u. Choosing $u = 0$ gives the result.

To establish the quadratic reciprocity law, the following Lemma is essential, which gives the evaluation of the quadratic Gauss sum

$$S(q, a) = \sum_{r=0}^{q-1} e^{2\pi i r^2 a/q}. \tag{5.52}$$

Lemma 5.2. (Landsberg-Schaal identity) *For any two coprime integers* $p, q > 0$

$$\frac{1}{\sqrt{p}} S(q, p) = \frac{e^{\pi i/4}}{\sqrt{2q}} \sum_{r=0}^{2q-1} e^{-\pi i r^2 p/(2q)}. \tag{5.53}$$

In particular, for any odd integer q,

$$S(q, 1) = \sum_{r=0}^{q-1} e^{2\pi i r^2/q} = \varepsilon(q)\sqrt{q}, \tag{5.54}$$

where

$$\varepsilon(q) = \begin{cases} 1 & \text{if } q \equiv 1 \bmod 4, \\ i & \text{if } q \equiv 3 \bmod 4. \end{cases} \tag{5.55}$$

Exercise 74. Suppose q is an odd prime, $\chi(\cdot) = \left(\frac{\cdot}{q}\right)$ the Legendre symbol and $(a, q) = 1$. Then prove that

$$S(q, a) = G_a(\chi). \tag{5.56}$$

Solution. Since $\chi(b) = \pm 1$ according as $b \equiv j^2 \pmod{q}$ or not, it follows that

$$\sum_{\substack{b=1 \\ (b,q)=1}}^{q} (1 + \chi(b))e^{2\pi i ba/q} = 2 \sum_{j=1}^{\frac{q-1}{2}} e^{2\pi i j^2 a/q} = \sum_{j=1}^{q-1} e^{2\pi i j^2 a/q}, \tag{5.57}$$

where the last equality follows from the fact that $(q - j)^2 \equiv j^2 \pmod{q}$.

Subtracting $\sum_{j=1}^{q-1} e^{2\pi i ba/q} = -1$ from the respective side of (5.56), we obtain (5.56).

Exercise 75. For $(a, b) = 1$ prove the multiplicativity

$$S(ab, 1) = S(a, b)S(b, a). \tag{5.58}$$

Solution. By the Chinese remainder theorem, every residue class $j \pmod{q}$ can be written as $bj_1 + aj_2$, $0 \le j_1 \le a - 1$, $0 \le j_2 \le b - 1$. Hence

$$S(ab, 1) = \sum_{j_1=1}^{a-1} \sum_{j_2=1}^{b-1} e^{2\pi i \frac{(bj_1 + aj_2)^2}{ab}}. \tag{5.59}$$

On expanding the square, we see that $S(ab, 1)$ is the product of $S(a, b)$ and $S(b, a)$, completing the proof.

Lemma 5.3. *Let q be an odd prime. Then for $(a, q) = 1$,*

$$S(q, a) = \left(\frac{a}{q}\right)\varepsilon(q)\sqrt{q}. \tag{5.60}$$

Proof.   By the separability (5.37), we have $G_a(\chi) = \chi(a)\tau(\chi)$. Substituting (5.56) and (5.54), we conclude (5.60), completing the proof.   □

Theorem 5.6. (Gauss quadratic reciprocity) *For distinct odd primes p, q, we have*

$$\left(\frac{p}{q}\right)\left(\frac{q}{p}\right) = (-1)^{\frac{p-1}{2}}(-1)^{\frac{q-1}{2}}. \tag{5.61}$$

Proof.   By (5.60),

$$S(pq, 1) = \varepsilon(pq)\sqrt{pq}$$

whose left-hand side is, by (5.58), $S(p, q)S(q, p)$. Hence substituting from (5.60), we obtain

$$S(pq, 1) = \left(\frac{p}{q}\right)\varepsilon(q)\sqrt{q}\left(\frac{q}{p}\right)\varepsilon(p)\sqrt{p}.$$

Comparing these, we conclude that

$$\left(\frac{p}{q}\right)\left(\frac{q}{p}\right)\varepsilon(q)\varepsilon(p) = \varepsilon(pq), \tag{5.62}$$

which amounts to (5.61), completing the proof.   □

It remains to prove Lemma 5.2 by means of (5.46).

We need the asymptotic formula for $\theta(x)$ when x is near the rational multiple of i. First we prepare

Exercise 76.

$$\sum_{n=-\infty}^{\infty} e^{-(b+nq)^2\varepsilon} = \frac{\sqrt{\pi}}{q\sqrt{\varepsilon}} + O(|\varepsilon|^2) \sim \frac{\sqrt{\pi}}{q\sqrt{\varepsilon}} \tag{5.63}$$

as $\varepsilon \to 0$ through $\operatorname{Re}\varepsilon > 0$.

This follows from the Euler-Maclaurin formula.

Proof.   (Proof of Lemma 5.2) Classifying the values of n modulo p, we find by Exercise 76 that

$$\theta\left(\varepsilon - \frac{2q}{p}i\right) = \sum_{a=0}^{p-1} e^{-\pi n^2(\varepsilon - \frac{2a}{p}i)} \sum_{n \equiv a \bmod p} e^{-\pi n^2\varepsilon} \sim \frac{\sqrt{\pi}S(p, q)}{p\sqrt{\varepsilon}}. \tag{5.64}$$

Now by (5.46), the left-hand side of (5.64) is

$$\left(\varepsilon - \frac{2q}{p}i\right)^{\frac{1}{2}}\theta\left(\frac{1}{\varepsilon - \frac{2q}{p}i}\right) = \left(\delta + \frac{p}{2q}i\right)^{\frac{1}{2}}\theta\left(\delta - \frac{-2p}{4q}i\right), \qquad (5.65)$$

where $\delta = \left(\varepsilon - \frac{2q}{p}i\right)^{\frac{1}{2}} - \frac{p}{2q}i$ and $\delta \to 0$ within $\operatorname{Re}\delta > 0$ as with ε. Hence the right-hand side of (5.65) is asymptotic to

$$\left(\delta + \frac{p}{2q}i\right)^{\frac{1}{2}}\frac{\sqrt{\pi}S(4q, -2p)}{4q\sqrt{\delta}}.$$

Since $\left(\delta + \frac{p}{2q}i\right)^{\frac{1}{2}} \sim \sqrt{\frac{ip}{2q}}$, it follows that

$$\frac{\sqrt{\pi}S(p,q)}{p\sqrt{\varepsilon}} \sim \sqrt{\frac{ip}{2q}}\frac{\sqrt{\pi}S(4q, -p)}{4q\sqrt{\delta}}$$

or

$$\sqrt{\frac{\delta}{\varepsilon}}\frac{S(p,q)}{p} = \sqrt{\frac{ip}{2q}}\frac{S(4q, -p)}{4q}. \qquad (5.66)$$

Since $\sqrt{\frac{\delta}{\varepsilon}} \sim \sqrt{\frac{-ip}{2q}}$, we conclude that

$$\sqrt{-i}\frac{S(p,q)}{\sqrt{p}} = \frac{S(4q, -p)}{2\sqrt{2q}}. \qquad (5.67)$$

We find that $\sqrt{-i} = e^{-\frac{\pi i}{4}}$ by setting $p = q = 1$ in (5.67). Hence noting that $S(4q, -p)$ reduces to $2\sum_{r=0}^{2q-1}e^{-\pi i r^2 p/(2q)}$ and this leads to (5.53), completing the proof. $\qquad\square$

5.3 Lambert series and character sums

In most of the existing literature, only the results are stated for L-functions with primitive characters because they take the simplest form. There are a few exceptions including Hasse [Hass], Joris [Jori], Neukirch [Neuk]. It has turned out that in order to utilize the group structure of characters, we need to treat all the characters, primitive and imprimitive alike. For this reason, it may be worth while stating basic facts about Dirichlet L-functions with imprimitive characters, too. In this section and in §5.5, we shall make clear that the Szmidt-Urbanowicz-Zagier [SUZa] method is in spirit exactly the same as the classical method for obtaining the values of the Riemann zeta-function at even positive integral arguments from the partial fraction expansion for the hyperbolic cotangent function $\coth x$ (or the cotangent function $\cot x$), which is a form of the Lambert series, and comparing the Laurent coefficients. Indeed, the following equality is the most well-known. We recall the proof in [Vist, Exercise 5.4] (cf. also [Böhh, pp. 39-43]).

Proposition 5.1. *Euler's identity*

$$\frac{B_{2m}}{(2m)!} = (-1)^{m-1} \frac{2\,\zeta(2m)}{(2\pi)^{2m}}, \quad m \geq 1 \tag{5.68}$$

follows from the partial fraction expansion for the hyperbolic cotangent function

$$\frac{1}{2}\coth \pi x = \frac{1}{e^{2\pi x} - 1} + \frac{1}{2} = \frac{1}{2\pi x} + \frac{x}{\pi}\sum_{n=1}^{\infty}\frac{1}{n^2 + x^2}, \quad \mathrm{Re}\,x \geq 0. \tag{5.69}$$

Proof. Rewriting (5.69) in the form

$$\frac{2\pi x}{e^{2\pi x} - 1} + \pi x = 1 + 2x^2\sum_{n=1}^{\infty}\frac{1}{n^2 + x^2} \tag{5.70}$$

and putting $2\pi x = z$, we obtain

$$\frac{z}{e^z - 1} = 1 - \frac{1}{2}z + 2z^2\sum_{n=1}^{\infty}\frac{1}{z^2 + (2\pi n)^2}, = 1 - \frac{1}{2}z + 2z^2\varphi\left(z^2\right)$$

say, where

$$\varphi(w) = \sum_{n=1}^{\infty}\left(w + 4\pi^2 n^2\right)^{-1}. \tag{5.71}$$

By Exercise 77 below, we obtain

$$\frac{z}{e^z - 1} = 1 - \frac{1}{2}z + \sum_{m=1}^{\infty}\frac{2\,(-1)^{m-1}}{(2\pi)^{2m}}\,\zeta(2m)\,z^{2m}, \quad |z| < 2\pi, \tag{5.72}$$

i.e. the zeta-value coefficients. On the other hand, the left-hand side has the expansion ((5.16))

$$\frac{z}{e^z - 1} = 1 - \frac{1}{2}z + \sum_{m=1}^{\infty}\frac{B_{2m}}{(2m)!}\,z^{2m}, \quad |z| < 2\pi. \tag{5.73}$$

Comparing the coefficients, we conclude (5.68). $\square$

Exercise 77. Expand $\varphi(w)$ into the Taylor series.

Solution. Since

$$\varphi^{(r)}(w) = (-1)^r r! \sum_{r=0}^{\infty} \left(w + 4\pi^2 n^2 \right)^{-r-1},$$

we see that

$$\frac{\varphi^{(r)}(0)}{r!} = (-1)^r \sum_{n=1}^{\infty} \frac{1}{(2\pi n)^{2r+2}} = \frac{(-1)^r}{(2\pi)^{2r+2}} \zeta(2r+2).$$

Hence

$$\frac{z}{e^z - 1} = 1 - \frac{1}{2} z + 2 \sum_{r=0}^{\infty} \frac{(-1)^r}{(2\pi)^{2r+2}} \zeta(2r+2) z^{2r+2}.$$

Here we also note that the partial fraction expansion for the hyperbolic cotangent function is equivalent to the functional equation (5.11) for the Riemann zeta-function, whose asymmetric form reads

$$\zeta(1-s) = 2^{1-s} \pi^{-s} \Gamma(s) \cos\left(\frac{\pi s}{2}\right) \zeta(s), \qquad (5.74)$$

which is a consequence of the Hurwitz formula (the functional equation for the Hurwitz zeta-function), which in turn is a consequence of the functional equation (5.15) for the Lipschitz-Lerch transcendent as stated prior to (5.15).

Remark 5.2. The most informative bibliography on Bernoulli numbers (and polynomials) is [DSSi] in which, however, the editors refer to [Apo1] but not [Apo2]. The same thing occurs in [SrCh], where there is an exposition of $\mathcal{B}_n(n, \alpha)$. In [Apo2], Apostol states that Carlitz has pointed out the source of these polynomials in [Eule], [Frob] and [Vand]. Thus these polynomials may be called "Eulerian polynomials" and are closely related to Mirimanoff polynomials (cf. [Vand]).

Throughout in what follows, whenever we refer to a character, we mean a Dirichlet character not necessarily primitive.

Let the modulus q be fixed throughout. For a Dirichlet character χ modulo q, let $L_\chi(t)$ denote the **Lambert series** associated to χ:

$$L_\chi(t) = \sum_{n=1}^{\infty} \chi(n) e^{-nt}, \quad \operatorname{Re} t > 0, \qquad (5.75)$$

which corresponds to the hyperbolic cotangent function above.

Szmidt-Urbanowicz-Zagier [SUZa, p. 275] deduced the expression

$$\sum_{0<n<\frac{N}{r}} \chi(n)e^{-rnt} = L_\chi(rt) - \frac{\bar{\chi}(r)}{\varphi(r)}e^{-Nt} \sum_{\psi \bmod r} \overline{\psi}(-N)L_{\chi\psi}(t). \qquad (5.76)$$

Thus we have the correspondence:

- the Taylor expansion of the right-hand side corresponds to (5.18) ((5.72)).

- the expansion of the left-hand side corresponds to (5.73) giving the weighted character sum $S_{u,r}^\kappa(\chi) = S_{r,N}^\kappa(\chi)$ defined by (5.78).

We fix positive integers N, r satisfying the conditions

$$N = uq, \ (r, N) = 1, \qquad (5.77)$$

and we introduce the notation

$$S_{u,r}^\kappa(\chi) = S_{r,N}^\kappa(\chi) = \frac{1}{q^\kappa} {\sum_{1\le a\le \frac{N}{r}}}' \chi(a)a^\kappa = \frac{1}{q^\kappa} {\sum_{1\le a\le \frac{u}{r}q}}' \chi(a)a^\kappa, \qquad (5.78)$$

the prime on the summation sign means that for the extremal value $\frac{N}{r} = \frac{uq}{r}$ of a, the corresponding summand is to be halved. In the notation of Yamamoto [Yama, p. 280], this is $S_{\frac{u}{r}} = S_{N/q}$. The essential case is $u \le r$, which we so assume.

Comparing the coefficients of both sides, they deduced

$$nr^{n-1} \sum_{0<a<\frac{uq}{r}} \chi(a)a^{n-1} = -B_{n,\chi}r^{n-1} + \frac{\bar{\chi}(r)}{\varphi(r)} \sum_\psi \overline{\psi}(uq)B_{n,\chi\psi}(-uq).$$
$$(5.79)$$

Substituting (5.21) and noting that $B_{0,\chi} = 0$ (χ non-trivial), we immediately obtain

Theorem 5.7. (Szmidt-Urbanowicz-Zagier) *Assume that u, r are positive integers and χ is a Dirichlet character modulo q such that $(uq, r) = 1$. Then*

$$S_{u,r}^\kappa(\chi) = \frac{1}{q^\kappa} {\sum_{1\le a\le \frac{u}{r}q}}' \chi(a)a^\kappa = -\frac{1}{q^\kappa}\frac{B_{\kappa+1,\chi}}{\kappa+1} + \frac{\bar{\chi}(r)}{(qr)^\kappa\varphi(r)} \sum_\psi \overline{\psi}(-uq)$$
$$(5.80)$$

$$\times \sum_{k=1}^{\kappa+1} \frac{1}{\kappa+1}\binom{\kappa+1}{k}(uq)^{\kappa+1-k}B_{k,\chi\psi},$$

where the sum is over all Dirichlet characters ψ modulo N.

Or

Theorem 5.8. *Assume that u, q, r are positive integers such that $(uq, r) = 1$ and χ is a Dirichlet character modulo q with conductor f and corresponding primitive character χ_1. Then we have*

$$S_{u,r}^{\kappa}(\chi) = -\frac{(-1)^{\kappa+1} + \chi(-1)}{2(\kappa+1)q^{\kappa}} \frac{(-1)^{\kappa}(\kappa+1)! f^{\kappa+1}}{(2\pi i)^{\kappa+1}\tau(\chi_1)} \prod_{p|q}(1 - \chi_1(p)p^{\kappa}) \quad (5.81)$$

$$\times L(\kappa+1, \chi_1) + \frac{\bar{\chi}(N)}{(qN)^{\kappa}\varphi(N)} \sum_{\psi} \bar{\psi}(-uq) \sum_{k=1}^{\kappa+1} \frac{1}{\kappa+1}\binom{\kappa+1}{k}(uq)^{\kappa+1-k}$$

$$\times \frac{(-1)^k + \chi\psi(-1)}{2} \frac{(-1)^{k-1}k! f'^k}{(2\pi i)^k \tau((\chi\psi)_1)} \prod_{p|qN}\left(1 - (\chi\psi)_1(p)p^{k-1}\right) L(k, (\chi\psi)_1)$$

where the sum is over all Dirichlet characters ψ modulo N and $\chi\psi$ is induced from $(\chi\psi)_1$ (mod f').

(5.81) follows from Lemma 5.4 below. Before stating it, we introduce the SUZ generalized Bernoulli numbers $B_{n,\chi}^{[r]}$. For any $0 \neq r \in \mathbb{Z}$, let

$$B_{n,\chi}^{[r]} = \prod_{p|r, p \text{ prime}} \left(1 - \chi(p)p^{n-1}\right) \cdot B_{n,\chi}. \quad (5.82)$$

Note that $B_{n,\chi}^{[r]} = B_{n,\chi}$ and that $B_{n,\chi}^{[r]} = B_{n,\chi}^{[r]}(0)$ is just $B_{n,\chi'}$ for the character χ' modulo $q|r|$ induced by χ, where $B_{n,\chi}^{[r]}(x)$ is the **SUZ generalized Bernoulli polynomial** defined by

$$B_{n,\chi}^{[r]}(x) = \sum_{k=0}^{n} \binom{n}{k} B_{k,\chi}^{[r]} x^{n-k}, \quad (5.83)$$

which has the properties $B_{n,\chi}^{[r]}(-x) = (-1)^n \chi(-1) B_{n,\chi}^{[r]}(x)$ unless $q = n = r = 1$, therefore $B_{n,\chi} = 0$ if $n > 1, \chi(-1) = (-1)^{n-1}$.

Lemma 5.4. (Szmidt-Urbanowicz-Zagier) *Assume that $f > 1$ indicates the conductor of Dirichlet character χ modulo q, χ_1 the corresponding primitive character of χ. Then we have*

$$B_{n,\chi_1}^{[q]} = B_{n,\chi} \quad (5.84)$$

$$= \begin{cases} 0 & \text{if } \chi(-1) = (-1)^{n-1} \\ \frac{(-1)^{n-1}n! f^n}{(2\pi i)^n \tau(\chi_1)} \prod_{p|q}\left(1 - \chi_1(p)p^{n-1}\right) L(n, \chi_1) & \text{if } \chi(-1) = (-1)^n. \end{cases}$$

Proof.   The proof follows from [SUZa, pp. 274-275]

$$B_{n,\chi} = B_{n,\chi_1} \prod_{p|q} \left(1 - \chi_1(p)p^{n-1}\right) = B_{n,\chi_1}^{[q]} \qquad (5.85)$$

and the properties of $B_{n,\chi}$ to the effect that $B_{n,\chi} = 0$ if $n > 1, \chi(-1) = (-1)^{n-1}$ while they give Dirichlet L-function values $L(n,\chi)$ at positive integers n if $\chi(-1) = (-1)^n$ as provided by (5.23).   □

The main result in [SUZa] is the divisibility result for the sums involving the $B_{n,\chi_1}^{[M]}$. It has turned out that (5.79) is quite useful in deriving a rich class of congruences in a very lucid fashion and there will appear several papers in this direction including [KUWa], [KUW1] subsequently.

5.4   Short interval character sums

In addition to the paper of Berndt [Ber2] (cited in [SUZa]), we refer to relevant papers of Johnson and Mitchell [JoJm] and of Yamamoto [Yama]. The latter is little known but is one of the most complete work in this field. His results are limited to the case of primitive characters, but can be readily generalized by using the general evaluation of the Gauss sum $G_n(\chi)$ in Hasse [Hass, pp. 444-450] (cf. also Joris [Jori]), where we recall the Gauss sum defined by

$$G_n(\chi) = \sum_{a=1}^{q} \chi(a)e^{2\pi i \frac{n}{q} a}. \qquad (5.1)$$

Lemma 5.5. *We use the notation in Theorem 5.8 with ψ for the primitive character inducing the character χ and we write for $n \in \mathbb{N}, n_0 = \frac{n}{(q,n)}, q_0 = \frac{q}{(q,n)}$. Then we have*

$$G_n(\chi) = \begin{cases} 0, & \text{if } f \nmid q_0, \\ \frac{\varphi(q)}{\varphi(q_0)}\mu\left(\frac{q_0}{f}\right)\psi\left(\frac{q_0}{f}\right)\overline{\psi}(n_0)\tau(\psi) & \text{if } f \mid q_0. \end{cases} \qquad (5.86)$$

Yamamoto's result in the general case reads

Theorem 5.9. *Let $\chi, f, \psi, q, q_0, n_0, \mu, \varphi$ be the same as in Lemma 5.5. Then*

$$S_{u,r}^{\kappa}(\chi) = \sum_{k=1}^{\kappa+1} \frac{\kappa!\varphi(q)\left(\frac{u}{r}\right)^{\kappa-k+1}\tau(\psi)}{(2\pi i)^k(\kappa-k+1)!} \sum_{n=1, f|q_0}^{\infty} \mu\left(\frac{q_0}{f}\right)\frac{\overline{\psi}(n_0)}{\varphi(q_0)}\frac{b_k(n)}{n^k}, \qquad (5.87)$$

where

$$b_k(n) = (-1)^{k+1}\chi(-1)\eta^{nu} - \eta^{-nu}, \quad (1 \leq k \leq \kappa),$$
$$b_{\kappa+1}(n) = (-1)^{\kappa+1}\chi(-1)(1 - \eta^{nu}) + 1 - \eta^{-nu}, \quad \eta = e^{\frac{2\pi i}{N}}. \tag{5.88}$$

We should be able to prove that Theorems 5.8 and 5.9 are equivalent on the ground that they both follow from the functional equation for the Riemann zeta-function.

We now state an equivalent form of Yamamoto's results in terms of Joris's evaluation of the Gauss sum. We denote the primitive character inducing χ to the modulus q by ψ with modulus f and introduce the notation

$$\tilde{q} = \prod_{\substack{p|q \\ p\nmid f}} p, \quad R = \frac{q}{f\tilde{q}}. \tag{5.89}$$

Lemma 5.6.

$$G_n(\chi) = \begin{cases} 0, & \text{if } R \nmid n, \\ \mu(\tilde{q})\varphi(\tilde{q})\tau(\psi)Rg\left(\frac{n}{R}\right) & \text{if } R \mid n, \end{cases} \tag{5.90}$$

where

$$g(n) = \mu((n,\tilde{q}))\varphi((n,\tilde{q}))\bar{\psi}(n). \tag{5.91}$$

Theorem 5.10. *Notation being the same as in Theorem 5.9, we have*

$$S_{u,r}^\kappa(\chi) = R\mu(\tilde{q})\psi(\tilde{q})\sum_{k=1}^{\kappa+1}\frac{\kappa!\varphi(q)\left(\frac{u}{N}\right)^{\kappa-k+1}\tau(\psi)}{(2\pi i)^k(\kappa-k+1)!} \tag{5.92}$$
$$\times \sum_{\substack{n=1 \\ R|n}}^{\infty}\mu((n,\tilde{q}))\varphi((n,\tilde{q}))\bar{\psi}(n)\frac{b_k(n)}{n^k},$$

where b_k are given by (5.88).

Exercise 78. For $p \equiv 1 \pmod 4$ prove Dirichlet's result.

$$S_{1/1,0}(\chi_{-4p}) = -2S_{1/4,0}(\chi_p). \tag{5.93}$$

Solution. Since

$$S_{1/1,0}(\chi_{-4p}) = \sum_{a=1}^{4p}\chi_4(a)\left(\frac{a}{p}\right), \tag{5.94}$$

we divide the sum into two parts $a \equiv 1 \pmod 4$ and $a \equiv -1 \pmod 4$ to obtain

$$S_{1/1,0}(\chi_{-4p}) = S_1 - S_{-1}, \tag{5.95}$$

where

$$S_{\pm 1} = \sum_{a=0}^{p-1} \left(\frac{4a \pm 1}{p} \right), \tag{5.96}$$

the sign being taken in the specified order.

We factor out $\left(\frac{4}{p} \right)$, which is 1 from S_1 to write

$$S_1 = \sum_{a=0}^{p-1} \left(\frac{a + \bar{4}}{p} \right),$$

where $\bar{4} = \frac{3p+1}{4} \equiv \frac{1-p}{4} \pmod{p}$ is the least positive residue of 4^{-1} modulo p.

Now writing the sum S_1 as

$$\sum_{a=0}^{\frac{p-1}{4}} + \sum_{a=1+\frac{p-1}{4}}^{p-1+\frac{p-1}{4}} - \sum_{a=p}^{p-1+\frac{p-1}{4}},$$

to obtain

$$S_1 = \sum_{a=0}^{\frac{p-1}{4}} \left(\frac{a + \bar{4}}{p} \right) + \sum_{a=1+\frac{p-1}{4}}^{p-1+\frac{p-1}{4}} \left(\frac{a + \bar{4} - p}{p} \right) - \sum_{a=p}^{p-1+\frac{p-1}{4}} \left(\frac{a + \bar{4}}{p} \right).$$

The first two terms sum to $\sum_{a=0}^{p-1} \left(\frac{a+\bar{4}}{p} \right)$, which is 0 and the last term is $-\sum_{a=1}^{\frac{p-1}{4}} \left(\frac{a+\frac{1-p}{4}}{p} \right)$, which after change of variable, becomes $-S_{1/4,0}(\chi_p)$.

Similarly, we may prove $S_{-1} = -S_{1/4,0}(\chi_p)$ proving the assertion.

The following formula is due to Dirichlet (cf. [BShh, p. 346]) and proved in a general form by Funakura [Funa]. We give a proof of a more general Theorem 5.11 below following Funakura.

Proposition 5.2. *Let* $k = \mathbb{Q}(\sqrt{d})$, $d < -2$ *and suppose the conductor* f_k *of* k *is odd. Then*

$$h_k = \frac{1}{2 - \left(\frac{d}{2} \right)} \sum_{\substack{0 < a < f_k/2 \\ (a, f_k) = 1}} \left(\frac{d}{a} \right), \tag{5.97}$$

where $\left(\frac{d}{a} \right)$ *is the Kronekcer symbol.*

Example 5.2. For $k = \mathbb{Q}(\sqrt{-5})$ we have $h_k = 2$. Hence $\mathcal{O}_k$ is not UFD. An example is $6 = 2 \cdot 3 = (1 + \sqrt{5})(1 - \sqrt{5})$.

Proof. Proposition 5.2 reads in this case

$$h_k = \frac{1}{2 - \chi_{-4}(2)} \sum_{a=1,3,7,9} \chi_{-4}(a) \left(\frac{a}{5}\right), \tag{5.98}$$

where χ_{-4} is the real primitive odd character to the modulus 4: $\chi_{-4}(-1) = -1$. Hence

$$h_k = \frac{1}{2} \left(1 + \chi_{-4}(-1)\left(\frac{-2}{5}\right) + \chi_{-4}(-1)\left(\frac{2}{5}\right) + \chi_{-4}(1)\left(\frac{2}{5}\right)\right) = 2 \tag{5.99}$$

because $\left(\frac{-1}{5}\right) = (-1)^{\frac{5-1}{2}} = 1$ and $\left(\frac{2}{5}\right) = (-1)^{\frac{5^2-1}{8}} = -1$. $\square$

Theorem 5.11. ([Funa, Theorem 10])

$$(1 - 2\chi(2)) \sum_{a=1}^{N} a\chi(a) = \chi(2)N \sum_{a=1}^{\left[\frac{N}{2}\right]} \chi(a),$$

i.e.

$$(1 - 2\chi(2))S_{1/1,1} = \chi(2)NS_{1/2,0}. \tag{5.100}$$

Proof. We note that (5.102) reads for a non-principal Dirichlet character mod N, N being odd,

$$\sum_{a=1}^{N} (-1)^a \chi(a) = \chi(2)2^{j+1}S_{1/2,0}(\chi) - S_{1/1,j}(\chi),$$

which is the equality in the proof of Funakura. The rest of the proof uses the finite Fourier transform and we refer to [Funa]. $\square$

Lemma 5.7. *For any function f, we have*

$$\sum_{a=1}^{N} (-1)^a f(2a) = 2 \sum_{a=1}^{\left[\frac{N}{2}\right]} f(2a) - \sum_{a=1}^{N} f(a), \tag{5.101}$$

and in particular for any Dirichlet character χ and any integer $j \geq 0$, we have

$$\sum_{a=1}^{N} (-1)^a (2a)^j \chi(2a) = 2^j \left(\chi(4)2^{j+1}S_{1/2,j}(\chi) - \chi(2)S_{1/1,j}(\chi)\right). \tag{5.102}$$

Proof. Proof follows immediately on classifying the values of a mod 2. $\square$

Remark 5.3. The Dirichlet class number formula for the imaginary quadratic field in finite form reads (cf. [Dave, p. 53])

$$h(d) = -\frac{1}{|d|} S_{1/1,1}(\chi_d),$$ (5.103)

in view of

$$L(1, \chi_d) = -\frac{\pi}{|d|^{\frac{3}{2}}} S_{1/1,1}(\chi_d),$$ (5.104)

whose general form (for an odd character χ mod q) is

$$L(1, \chi) = -\frac{\pi i}{\tau(\chi) q} S_{1/1,1}(\chi).$$ (5.105)

Combining (5.100) and (5.104) gives rise to

$$S_{1/2,0}(\chi_d) = (2 - \chi_d(2)) h(d),$$ (5.106)

which is [Ber2, Corollary 3.4].

The 1/4-th sum is also due to Dirichlet ([Dick, Vol. 3, (5), (ii), p. 101]; cf. [Ber2, Theorem 3.7]).

$$h(-4d) = 2 S_{1/4,0}(\chi_d)$$ (5.107)

whose special case with $p \equiv 1$ mod 4 has been essentially used in Chowla [PCho, p. 58] and also in [ZhXu] to deduce from a form of the class number formula (5.104) their main ingredient in the disguised form

$$S_{1/4,0}(\chi_p) = \frac{\tau(\chi_p)}{\pi} L(1, \chi_4 \chi_p).$$ (5.108)

5.5   Riemann-Hecke-Bochner correspondence and character sums

Our aim in this section is, in view of the correspondence referred to at the beginning of §5.3, to give a functional equational proof of (5.79) (or rather its counterpart). This philosophy is based on the Riemann-Hecke-Bochner correspondence [Knop]. Combining [Jori] and [Neuk, pp. 211-215], we see that the essential ingredient is the ℓ-function defined by (5.13).

Taking the Mellin transform of both sides of (5.76), we obtain

$$\Gamma(s) \sum_{0 < n < \frac{N}{r}} \frac{\chi(n)}{(rn)^s} = \Gamma(s) \sum_{n=1}^{\infty} \frac{\chi(n)}{(rn)^s} - \Gamma(s) \frac{\overline{\chi}(r)}{\varphi(r)} \sum_{\psi} \overline{\psi}(-N)$$

$$\times (L(\chi\psi, N, s) - M_\chi(N)),$$

where $M_\chi(N) = \sum_{n=1}^{N-1} \frac{\chi\psi(n)}{n^s}$ is the weighted complete character sum and $L(\chi, x, s)$ is the Hurwitz-Lerch L-function introduced by Morita.

Theorem 5.12.

$$\sum_{0<n<\frac{N}{r}} \frac{\chi(n)}{n^s} = L(s, \chi) - \frac{\overline{\chi}(r)}{\varphi(r)} r^s \sum_\psi \overline{\psi}(-N)(L(\chi\psi, N, s) - M_\chi(N)).$$

$$(5.109)$$

Yamamoto's method gives not only a counterpart of the Schmidt-Urbanowics-Zagier formula but also also gives the general evaluation for ([Yama, p. 285])

$$T_{u,r}^\kappa(\chi) = T_{u,N}^\kappa(\chi) = \sum_{a=0}^{N-1} \chi(a)\tilde{f}\left(\frac{a}{N}\right), \qquad (5.110)$$

where $\tilde{f}$ is the conjugate function to the function $f(x)$, which is x^κ for $0 < x \le u/r$ and is 0 for $u/r < x \le 1$. This $f(x)$ is used in deriving Theorem 5.9 and has the Fourier coefficients $\hat{f}_n$. It has the Fourier series

$$\tilde{f}(x) \sim -i \sum_{n=-\infty}^\infty \text{sgn}(n)\hat{f}_n e^{2\pi inx} \qquad (5.111)$$

convergent to $\tilde{f}$ except for $x = 0$, u/r. (5.110) is a counterpart of the sum defined by (5.78).

Theorem 5.13. *Let* $\chi, f, \psi, q, q_0, n_0, \mu, \varphi$ *be the same as above, assume* $q \nmid r$. *Then*

$$T_{u,r}^\kappa(\chi) = -i \sum_{k=1}^{\kappa+1} \frac{\left(\frac{u}{N}\right)^{\kappa-k+1} \kappa! \varphi(q)\tau(\psi)}{(2\pi i)^{\kappa+1}(\kappa-k+1)!} \sum_{n=1, f|q_0}^\infty \mu\left(\frac{q_0}{f}\right) \psi\left(\frac{q_0}{f}\right) \frac{\overline{\psi}(n_0)}{\varphi(q_0)} \frac{\tilde{b}_k(n)}{n^k}$$

$$(5.112)$$

where

$$\tilde{b}_k(n) = (-1)^k \chi(-1)\eta^{nu} - \eta^{-nu}, \quad (1 \le k \le \kappa),$$

$$\tilde{b}_{\kappa+1}(n) = (-1)^\kappa \chi(-1)(1-\eta^{nu}) + 1 - \eta^{-nu}, \quad \eta = e^{\frac{2\pi i}{N}}.$$

$$(5.113)$$

Proof. We note that the sum for $T_{u,r}^\kappa(\chi)$ is taken over all $(a, q) = 1$, and it has the Fourier expansion in view of (5.111):

$$
T^{\kappa}_{u,r} = -i \sum_{a=0}^{q-1} \chi(a) \sum_{n=-\infty}^{\infty} \operatorname{sgn}(n) \hat{f}_n e^{\frac{2\pi i n a}{q}}
$$

$$
= -i \sum_{n=1}^{\infty} \left(\hat{f}_n - \chi(-1)\hat{f}_{-n} \right) G_n(\chi)
$$

$$
= -i \sum_{k=1}^{\kappa+1} \frac{\left(\frac{u}{N}\right)^{\kappa-k+1} \kappa!}{(2\pi i)^{\kappa+1}(\kappa-k+1)!} \sum_{n=1}^{\infty} \frac{G_n(\chi)\tilde{b}_k(n)}{n^k},
$$

where $\tilde{b}_k(n)$ is defined by (5.113). Substituting Lemma 5.6, we conclude (5.112), completing the proof.   □

It goes without saying that these short interval character sums were considered in relation to the evaluation of class numbers in the times without computers. Although from computational point of view only, short sums are not needed, the interest still continues of the relationship between short character sums and the class numbers. Yamamoto's paper [Yama] has been devoted to elucidating this relationship from a general point of view, unifying the previous results in [JoJm], [Ber2]. For the most recent references, cf. [SUWa].

5.6   The ℓ-function

Theorem 5.14. (Kubert-Lang [KuLa, Chapter 1, §2]) *For all* $s \in \mathbb{C} - \{1\}$, *we have*

$$
q^{s-1}\ell(s,\chi) = \prod_{p|\tilde{q}} \left(1 - \psi(p)p^{s-1}\right) f^{s-1}\ell(s.\psi). \tag{5.114}
$$

Proof.   Recall the Kubert identity satisfied by $l_s(x)$:

$$
l_s(x) = m^{s-1} \sum_{a=0}^{m-1} l_s\left(\frac{x+a}{m}\right), \quad \forall m \in \mathbb{N}. \tag{5.115}
$$

Classifying residues mod q with respect to mod f, we obtain

$$
\ell(s,\chi) = \sum_{\substack{b\in\mathbb{Z}/f\mathbb{Z}}} \sum_{\substack{a\in\mathbb{Z}/q\mathbb{Z} \\ a\equiv b(f)}} \chi(c) l_s\left(\frac{c}{q}\right).
$$

If $\tilde{q} = 1$, then $(a, q) = 1$ and $(a, f) = 1$ are the same, and we have $\chi(a) = \psi(a)$, $(a, f) = 1$,

$$\ell(s, \chi) = \sum_{b \in \mathbb{Z}/f\mathbb{Z}} \psi(b) \sum_{a \in \mathbb{Z}/q\mathbb{Z}} l_s \left(\frac{b + \frac{a}{f}}{q/f} \right),$$

the inner sum being $(q/f)^{1-s}$ whence (5.114) follows. If $\tilde{q} \neq 1$, say $q = fp$, with a prime $p, (p, f) = 1$, then $\chi(a) = \psi(a)$, for $(a, f) = 1$, but $\chi(c) = 0$ if $a = pc$, $c \in \mathbb{Z}$. Hence,

$$\ell(s, \chi) = \sum_{b \in \mathbb{Z}/f\mathbb{Z}} \psi(b) \sum_{\substack{a \in \mathbb{Z}/q\mathbb{Z} \\ a \equiv 1(f)}} l_s \left(\frac{b + \frac{a}{f}}{q/f} \right) - \sum_{\substack{a \in \mathbb{Z} \\ a \equiv b(f)}} \psi(a) l_s \left(\frac{a}{q} \right).$$

The first sum on the right is the sum as before and the second is

$$\psi(p) \sum_{c \in \mathbb{Z}/f\mathbb{Z}} \psi(c) l_s \left(\frac{a}{q} \right) = \psi(p) \ell(s, \psi).$$

Hence it follows that

$$\ell(s, \chi) = \left(\frac{q}{f} \right)^{1-s} \ell(s, \psi) - \psi(p) \ell(s, \psi),$$

which is (5.115) with $q = fp$.

The case where the above $\tilde{q}$ contains more prime factors, is treated similarly, completing the proof. $\qquad\square$

Theorem 5.15. (Neukirch [Neuk, Proposition 4.2]) *We write the primitive character inducing χ to the modulus q by ψ with modulus f. For all $s \in \mathbb{C} - \{1\}$, we have*

$$L(s, \bar{\chi}) = g(s, \chi) \ell(s, \chi), \tag{5.116}$$

where

$$g(s, \chi) = \frac{1}{\tau(\psi)} \left(\frac{q}{f} \right)^{s-1} \prod_{\substack{p|q \\ p \nmid f}} \frac{1 - \bar{\psi}(p) p^{-s}}{1 - \psi(p) p^{s-1}}. \tag{5.117}$$

Proof. Both sides of (5.116) being analytic, it suffices to prove it in the case $\sigma > 1$. First consider the case χ being primitive. Then (5.37) is valid. Hence dividing both sides by n^s and summing over all $n \in \mathbb{N}$, we obtain $(\sigma > 1)$

$$\tau(\chi) L(s, \bar{\chi}) = \sum_{a \bmod q} \chi(a) \ell_s \left(\frac{a}{q} \right) = \ell(s, \chi). \tag{5.118}$$

In the general case of χ we appeal to (1.43) (Chapter 1) and apply (5.118) in the form

$$L(s,\bar{\chi}) = \frac{1}{\tau(\psi)} \prod_{\substack{p|q \\ p\nmid f}} (1 - \bar{\psi}(p)p^{-s})\ell(s,\psi). \qquad (5.119)$$

Then the conclusion follows from Theorem 5.14.   □

Remark 5.4. (i) In [Neuk] $\tau(\psi)$ in (5.119) is written as $\tau(\chi)$. With this typo corrected, [Neuk, Proposition 4.2] (=(5.116)) and

$$\ell(s,\chi) = R^{1-s}\tau(\psi)\mu(q)\psi(q)L(s.\bar{\psi}) \prod_{p|q} \left(1 - \bar{\psi}(p)p^{1-s}\right), \qquad (5.120)$$

which is [Jori, (7)] amount to the same thing, where $\tilde{q}$, R are defined by (5.89).

The link between (5.120) and (5.116) is

$$\prod_{p|\tilde{q}} \left(1 - \psi(p)p^{-\alpha}\right) = \tilde{q}^{-\alpha}\mu(\tilde{q})\psi(\tilde{q}) \prod_{p|\tilde{q}} \left(1 - \bar{\psi}(p)p^{\alpha}\right). \qquad (5.121)$$

(ii) It is interesting to notice that the several relevant papers [Ber2], [JoJm], [Jori] and [Yama] have appeared around the same year 1976 and only [Ber2] has been most well-known and most frequently cited. However, as demonstrated above, [Yama] supersedes all others.

5.7   Discrete mean square results

The discrete mean value of the special values of the Dirichlet L-function $L(s,\chi)$—especially that of $L(1,\chi)$ in view of its relevance to the class number of the associated number fields—has been the subject of many researches. One can consult an excellent survey of Matsumoto [Mats] for the reference and [KTYZ], where the $s = 1$ case has been completely and structurally settled. The discrete mean square at positive integers has been also considered extensively by several authors. Katsurada and Matsumoto [KaMa] were the first who obtained the result with unspecified coefficients. Louboutin [Lou1], [Lou2], [Lou3] considered the same problem and made the coefficients explicit. Liu and Zhang [LiuZ] made the coefficients more explicit than Louboutin, who also considered the cases of the product of two Dirichlet L-functions. Their result has been fully generalized by [KMZh].

However in none of these papers (save for [KTYZ]), attention is paid on the reason why the formula is to hold, i.e. the underlying structure that forces the formula to hold has never been studied and only ad-hoc methods have been adopted.

The underlying principle is exactly the same as in [BKTS] if we use another basis (Hurwitz zeta-function) for relevant periodic functions, as elucidated by [HKTo], and then that the characteristic difference properties of the Hurwitz zeta-function will show its essential effect and just telescoping gives the result, as in infinitesimal calculus–differentiation and integration!

In this section, to show some historically interesting feature of the problem, we treat the case of negative integers in terms of Bernoulli polynomials although this case could be included in the positive integer case in terms of the Hurwitz zeta-function through (5.122). We get an interesting convolution identity as a bonus.

Notation

$L(s, \chi)$ is the Dirichlet L-function associated with a Dirichlet character χ to the modulus q defined by (5.3). We assume throughout that $q \geq 3$.

$B_k(x)$ is the k-th Bernoulli polynomial define by (3.35) for which we have (5.20).

$B_k = B_k(0)$ is the k-th Bernoulli number defined by (3.36).

$\overline{B}_k(x)$ is the periodic Bernoulli polynomial introduced at the end of §3.2 Chapter 3:

$$\overline{B}_k(x) = B_k(x - [x])$$

with $[x]$ designating the integral part of x.

We note for a non-negative integer n,

$$\zeta(-n, x) = -\frac{1}{n+1} B_{n+1}(x) \tag{5.122}$$

([Vist, Chapter 4]).

$$J_k(q) = \sum_{d \mid q} \mu\left(\frac{q}{d}\right) d^k \tag{5.123}$$

is the Jordan totient function, where the summation is extended over all positive divisors of q and μ is the Möbius function. Note that

$$J_1(q) = \varphi(q) = \sum{}^{*} 1 \tag{5.124}$$

is the Euler function, where $*$ on the summation sign means that it is extended over those natural numbers relatively prime to q.

We are in a position to state our results.

Theorem 5.16. *For a non-negative integer n, we have*

$$\sum_{\chi \bmod q} |L(-n,\chi)|^2 = \varphi(q)q^{2n}\frac{B_{2n+2}}{(n+1)^2}J_{-2n-1}(q)$$

$$+ \frac{2}{n+1}\varphi(q)q^{2n}\sum_{r=0}^{n-1}\binom{n+1}{r}\frac{B_{n-r+1}B_{n+r+1}}{n+r+1}J_{-n-r}(q)$$

$$+ (-1)^n\varphi^2(q)q^{2n}\frac{1}{(n+1)^2}\frac{B_{2n+2}}{\binom{2n+2}{n+2}}. \tag{5.125}$$

In conjunction with (5.131) below, the case $s = 0$ is of some interest:

$$\sum_{\chi \bmod q} |L(0,\chi)|^2 = \frac{1}{6}\varphi(q)J_{-1}(q) + \frac{1}{12}\varphi^2(q).$$

We note that if we appeal to the relation

$$\frac{1}{n+1}\binom{n+1}{r}\frac{1}{n+r+1}B_{n-r+1}B_{n+r+1} = \binom{n}{r}\frac{B_{n-r+1}}{n-r+1}\frac{B_{n+r+1}}{n+r+1}$$

$$= \binom{n}{r}\zeta(-n+r)\zeta(-n-r),$$

then (5.125) reads

$$\sum_{\chi \bmod q} |L(-n,\chi)|^2 = \frac{2}{n+1}\varphi(q)q^{2n}\zeta(-2n-1)J_{-2n-1}(q)$$

$$+ \varphi(q)q^{2n}\sum_{r=0}^{n-1}\binom{n}{r}\zeta(-n+r)\zeta(-n-r)J_{-n-r}(q)$$

$$+ \frac{2(-1)^n}{n+1}\varphi^2(q)q^{2n}\frac{1}{\binom{2n+2}{n+1}}\zeta(-2n-1), \tag{5.126}$$

which is Theorem 6 of Katsurada and Matsumoto [KaMa].

We state the positive integer case $s = k > 1$ treated by Katsurada and Motsumoto [KaMa], Louboutin [Lou2], Liu and Zhang [LiuZ] and Kanemitsu, Ma and Zhang [KMZh]. The following result amounts to the Katsurada-Motsumoto Theorem in the spirit of [KaMa].

Theorem 5.17. *For integers $k > 1$ we have*

$$\sum_{\chi \bmod q} |L(k, \chi)|^2 = \frac{\varphi(q)}{q^{2k}} J_{2k}(q)\zeta(2k)$$

$$+\frac{2\varphi(q)}{q^{2k}} \sum_{\substack{r=0 \\ r \neq k-1}}^{N} \binom{k+r-1}{r-1} \zeta(k+r)\zeta(k-r)J_{k-r}(q)$$

$$+\frac{2(-1)^{k-1}}{q^{2k}} \binom{2k-2}{k-1} \zeta(2k-1)\varphi(q)^2 \left(\log q + \sum_{p|q} \frac{\log p}{p-1} - \frac{\zeta'}{\zeta}(2k-1) + \gamma \right)$$

$$+\frac{\varphi(q)}{q^{2k}} R_N(q),$$

where γ is the Euler constant and $R_N(q)$ is defined by

$$R_N(q) = \sum_{d|q} \mu\left(\frac{q}{d}\right) \frac{1}{d^{N-k}} \sum_{m=1}^{\infty} \frac{1}{m^{k+N}} \int_1^\infty \overline{B}_{N-k+1}(dn+z)z^{r-N-1}\,dz.$$

$$(5.127)$$

5.8 Proof of Theorem 5.16

Exercise 79. For a character χ mod q and an arbitrary arithmetic function $f(a)$ let

$$B(\chi) = \sum_{a \bmod q} \chi(a)f(a). \tag{5.128}$$

Then prove that

$$S := \sum_{\chi} |B(\chi)|^2 = \varphi(q) \sum_{d|q} \mu\left(\frac{q}{d}\right) \sum_{a=1}^{d-1} \left| f\left(\frac{a}{d}\right) \right|^2. \tag{5.129}$$

Solution. Taking the square of moduli of both sides of (5.128) and expanding, we will encounter the sum

$$\sum_{\chi \bmod q} \chi(a)\bar{\chi}(b),$$

which is $\varphi(q)$ if $a \equiv b$ and $(b, q) = 1$ and 0 otherwise. Hence our sum S becomes

$$S = \sum_{\substack{a=1 \\ (a,q)=1}}^{q} \left| f\left(\frac{a}{q}\right) \right|^2.$$

Applying the relative primality principle in Remark 3.1, (i), we find that

$$S = \varphi(q) \sum_{d|q} \sum_{a'=1}^{q/d-1} \mu(a) \left| f\left(\frac{a'}{q/d}\right) \right|^2, \tag{5.130}$$

which amounts to (5.129). This completes the solution.

Applying Exercise 79 to Formula (5.8), we deduce

Lemma 5.8. *For $s \neq 1$,*

$$\sum_{\chi} |L(s,\chi)|^2 = \frac{\varphi(q)}{q^{2\sigma}} \sum_{d|q} \mu\left(\frac{q}{d}\right) \sum_{a=1}^{d} \left| \zeta\left(s,\frac{a}{d}\right) \right|^2. \tag{5.131}$$

Lemma 5.9. (Nielsen [N1, p.76, (10)]) *For each $n \geq 1$,*

$$B_n(x)^2 = B_{2n}(x) + 2n \sum_{r=0}^{n-2} \binom{n}{r} \frac{B_{n-r}B_{n+r}(x)}{n+r} \tag{5.132}$$

$$+ \frac{(-1)^{n-1}B_{2n}}{\binom{2n}{n}}.$$

Proof. We give an independent proof of Nielsen's, which helps to understand the characteristic difference property of the Bernoulli polynomials. We apply the method of undetermined coefficients. First we have the Bernoulli polynomial expansion

$$B_n(x)^2 = \sum_{k=1}^{2n} a_k B_k(x) + a_0, \tag{5.133}$$

where a_k's are to be determined ($a_k = a_k(n)$). To this end we compare the two expressions for

$$\Delta B_n^2(x) = B_n(x+1)^2 - B_n(x)^2.$$

On one hand, by the characteristic difference equation

$$B_n(x+1) - B_n(x) = nx^{n-1} \tag{5.134}$$

yields

$$\Delta B_n^2(x) = (B_n(x+1) - B_n(x))(B_n(x+1) + B_n(x))$$
$$= nx^{n-1}\left(nx^{n-1} + 2B_n(x)\right).$$

Hence by (5.20)

$$\Delta B_n^2(x) = 2nx^{2n-1} + 2n \sum_{r=0}^{n} \binom{n}{r} B_{n-r}x^{n+r-1}. \tag{5.135}$$

On the other hand, by (5.134) again,

$$\Delta B_n^2(x) = \sum_{k=1}^{2n} a_k \left(B_k(x+1) - B_k(x)\right) = \sum_{k=1}^{2n} a_k k x^{k-1},$$

or

$$\Delta B_n^2(x) = \sum_{k=1}^{n-1} k a_k x^{k-1} + \sum_{r=0}^{n}(n+r)a_{n+r}x^{n+r-1}. \tag{5.136}$$

Comparing the coefficients in (5.135) and (5.136), we conclude that

$$a_{2n} = 1, \quad a_{2n-1} = 0, \quad a_{n+r} = \frac{2n}{n+r}\binom{n}{r}B_{n-r}, 0 \le r \le n-2 \tag{5.137}$$

$$a_k = 0, \quad 1 \le k \le n-1,$$

which establishes (5.132) save for the value of $a_0 = a_0(n)$. For this we compute the Spannenintegral of $B_n(x)^2$:

$$\int_0^1 B_n(x)^2 \mathrm{d}x = \sum_{k=1}^{2n} a_k \int_0^1 B_k(x)\mathrm{d}x + \int_0^1 a_0 \mathrm{d}x = a_0,$$

by the orthogonality. We recall the value of the above integral (cf. e.g. [KTZh] or [Niel]):

$$(a_0(n) =) \int_0^1 B_n(x)^2 \mathrm{d}x = (-1)^{n-1}\frac{B_{2n}}{\binom{2n}{n}}, \qquad n = 1, 2, \cdots, \tag{5.138}$$

whence (5.132) follows. □

As a bonus we obtain the following interesting convolution formula:

Proposition 5.3. *For $n \ge 2$, we have*

$$B_n^2 - 2n\sum_{r=0}^{n}\binom{n}{r}B_{n-r}\frac{B_{n+r}}{n+r} = (-1)^{n-1}\frac{B_{2n}}{\binom{2n}{n}}.$$

Proof. This follows on equating the identity (the case $x = 0$ of (5.133))

$$B_n^2 = \sum_{k=1}^{2n} a_k B_k + a_0,$$

the values (5.137) of $a_k(n)$ and (5.138). □

Proof of Theorem 5.16. By (5.122),

$$S(d) := \sum_{a=1}^{d} \left| \zeta\left(-n, \frac{a}{d}\right) \right|^2 = \frac{1}{(n+1)^2} \sum_{a=1}^{d} B_{n+1}\left(\frac{a}{d}\right)^2,$$

which, by Lemma 5.9, becomes

$$S(d) = \frac{1}{(n+1)^2} \sum_{a=1}^{d} B_{2n+2}\left(\frac{a}{d}\right)$$

$$+ \frac{2}{n+1} \sum_{r=0}^{n-1} \binom{n+1}{r} \frac{B_{n-r+1}}{n+r+1} \sum_{a=1}^{d} B_{n+r+1}\left(\frac{a}{d}\right) + \frac{d}{(n+1)^2} a_0(n+1)$$

$$= \frac{1}{(n+1)^2} B_{2n+2} d^{-2n-1} + \frac{2}{n+1} \sum_{r=0}^{n-r+1} \binom{n+1}{r} B_{n-r+1} \frac{B_{n+r+1}}{n+r+1} d^{-n-r}$$

$$+ \frac{d}{(n+1)^2} a_0(n+1) \tag{5.139}$$

by the Kubert relation for the Bernoulli polynomial

$$B_s(dx) = d^{s-1} \sum_{a=1}^{d} B_s\left(x + \frac{a}{d}\right). \tag{5.140}$$

Now substituting (5.139) in the formula of Lemma 5.8, we complete the proof of Theorem 5.16.

Chapter 6

Control systems and number theory

In this chapter, we take up the discipline of control theory, which looks unrelated to number theory, trying to envisage the number-theoretic as well as mathematical aspects thereof. It is not quite surprising to treat such a subject in the number theory framework, as can be seen in the name of some schools, "School of mathematics and systems science".

Our primary concern is grasping some fundamental concepts in control theory from the number-theoretic point of view. Based partially on the first author's dissertation, we treat the chain scattering representation (of a plant) and H^∞-control problem. The homographic transformation of the former works as the action of the symplectic group on the Siegel upper half-space (in the case of constant matrices) and we will have a new look at the control as a group action. We hope the rich theory of Siegel modular forms will show its effect in control theory some time. A very interesting correspondence shows up here between the transfer functions in control theory and the zeta-functions in number theory. Especially, if we choose the Riemann zeta-function shifted by 1, $\zeta(s+1)$, where $s = \sigma + j\omega$, then we can see a very close correspondence between their region of stability and analyticity (both are $\sigma > 0$), the critical lines ($\sigma = 0$ and $\sigma = -\frac{1}{2}$, respectively), only the functional equation aspect being not clear in the former since the transfer functions are mostly just rational functions. Here, we might introduce a more sophisticated transfer function, as in the case of FOPID (Fractional Order Proportional-Integral-Differential) control. It turns out that they are the Riemann-Liouville fractional integral transform (6.76) of the input function.

On the other hand, the H^∞-norm or more generally, H^{2k}-norm problem appear as the estimate of effectiveness of control. In addition to the correspondence alluded to above, striking similarities appear between control theory and zeta-functions if we consider the power norm (6.55), where we can see a very close correspondence between $2k$-mean values of the zeta-function and the H^{2k} control problem.

We might expect that a finer theory of transfer functions be developed which would be certain special functions whose avatars appear as rational functions (e.g. Padè approximation thereof).

After a brief introduction, we go on to §6.2, where we expound the elements of linear systems from the point of view of our principle of visualization of the state, an interface between the past and the present. We view all the systems as embedded in the state equation, thus visualizing the state. Then we go on to §6.3 treating the chain scattering representation of the plant a là [Kimu], which includes the feedback connection in a natural way and we consider the H^∞-control problem in this framework. As stated in [Kimu, p. 9, p. 68], the main reason for using the chain scattering representation is that it represents the cascade connection in a very lucid way as the product. We may view in particular the unity feedback system as accommodated in the chain scattering representation, giving a better insight into the structure of the system. In the unity feedback system included is the *PID*-compensator whose generalization to fractional order calculus is the FOPID controller mentioned above.

Our main concern being the exhibition of similarities between control systems and number theory, we may state only fragments of control theory itself and some remain speculative (but hopefully more accessible to non-specialists). The interested reader may consult more specified books including [Haya], [HeMe], [Kimu], etc.

6.1   Introduction and preliminaries

It turns out that there is great similarity in control theory and number theory in their treatment of the signals in time domain (t) and frequency domain (ω) or an expanded frequency domain ($s = \sigma + j\omega$) which is conducted by the Laplace transform in the case of control theory while in the theory of zeta-functions, this role is played by the Mellin transform, both of which convert the signals in time domain to those in the right half-plane (expanded frequency domain). Following the tradition of electrical engineering, we use the symbol j to indicate the imaginary unit i hereafter. For integral transforms, cf. §6.9.

§6.5 introduces the Hardy space H_p which consists of functions analytic in $\mathcal{RHP}$—right half-plane $\sigma > 0$ and are in L^p at the boundary $\sigma = 0$.

6.2   State space representation and the visualization principle

Let $\mathbf{x} = \mathbf{x}(t) \in \mathbb{R}^n$, $\mathbf{u} = \mathbf{u}(t) \in \mathbb{R}^r$ and $\mathbf{y} = \mathbf{y}(t) \in \mathbb{R}^m$ be the **state** function, **input** function and **output** function, respectively. We write $\dot{\mathbf{x}}$ for $\frac{\mathrm{d}}{\mathrm{d}t}\mathbf{x}$. The system of (differential equations) DEs

$$\begin{cases} \dot{\mathbf{x}} = A\mathbf{x} + B\mathbf{u}, \\ \mathbf{y} = C\mathbf{x} + D\mathbf{u} \end{cases} \tag{6.1}$$

is called a **state equation** for a **linear system**, where $A \in M_{n,n}(\mathbb{R})$, B, C, D are given constant matrices.

The state $\mathbf{x}$ is not visible while the input and output are so, and the state may be thought of as an interface between the past and the present information since it contains all the information contained in the system from the past. The $\mathbf{x}$ being invisible, (6.1) would read

$$\mathbf{y} = D\mathbf{u}, \tag{6.2}$$

which appears in many places in literature in disguised form. All the subsequent systems e.g. (6.29) are variations of (6.2). And whenever we would like to obtain the state equation, we are to restore the state $\mathbf{x}$ to make a recourse to (6.1), which we would call the **visualization principle**. In the case of feedback system, it is often the case that (6.2) is given in the form of (6.36). It is quite remarkable that this controller S works for the matrix variable in the symplectic geometry (cf. §6.4).

Using the matrix exponential function e^{At}, the first equation in (6.1) can be solved in the same way as for the scalar case:

$$\mathbf{x} = \mathbf{x}(t) = e^{At}\mathbf{x}(0) + Be^{At}\int_0^t e^{-A\tau}\mathbf{u}(t)\,\mathrm{d}\tau. \tag{6.3}$$

Definition 6.1. A linear system with the input $\mathbf{u} = \mathbf{o}$

$$\dot{\mathbf{x}} = \frac{\mathrm{d}}{\mathrm{d}t}\mathbf{x} = A\mathbf{x}, \tag{6.4}$$

called an **autonomous system**, is said to be **asymptotically stable** if for all initial values, $\mathbf{x}(t)$ approaches a limit as $t \to \infty$.

Since the solution of (6.4) is given by

$$\mathbf{x} = e^{At}\mathbf{x}(0), \tag{6.5}$$

the system is asymptotically stable if and only if

$$||e^{At}|| \to 0 \quad \text{as} \quad t \to \infty. \tag{6.6}$$

A linear system is said to be **stable** if (6.6) holds, which is the case if all the eigenvalues of A have negative real parts. Cf. §6.5 in this regard. It also amounts to saying that the step response of the system approaches a limit as time elapses, where **step response** means a response

$$\mathbf{y}(t) = \int_0^t e^{A(t-\tau)} u(\tau) \, d\tau, \tag{6.7}$$

with the **unit step function** $u = u(t)$ as the input function, which is 0 for $t < 0$ and 1 for $t \geq 0$.

Up here, the things are happening in the time domain. We now move to a frequency domain. For this purpose, we refer to the Laplace transform to be discussed in §6.9. It has the effect of shifting from the time domain to frequency domain and vice versa. For more details, see e.g. [Kimu]. Taking the Laplace transform of (6.1) with $\mathbf{x}(0) = \mathbf{o}$, we obtain

$$\begin{cases} sX(s) = AX(s) + BU(s) \\ Y(s) = CX(s) + DU(s), \end{cases} \tag{6.8}$$

which we solve as

$$Y(s) = G(s)U(s), \tag{6.9}$$

where

$$G(s) = C(sI - A)^{-1}B + D, \tag{6.10}$$

where I indicates the identity matrix, which is sometimes denoted I_n to show its size.

In general, supposing that the initial values of all the signals in a system are 0, we call the ratio of output/input of the signal, the **transfer function**, and denote it by $G(s)$, $\Phi(s)$, etc. We may suppose so because if the system is in equilibrium, then we may take the values of parameters at that moment as standard and may suppose the initial values to be 0.

(6.10) is called the **state space representation** (form, realization, description, characterization) of the transfer function $G(s)$ of the system (6.1), and is written as

$$G(s) = \left(\begin{array}{c|c} A & B \\ \hline C & D \end{array} \right). \tag{6.11}$$

According to the visualization principle above, we have the **embedding principle**: Given a state space representation of a transfer function $G(s)$, it is to be embedded in the **state equation** (6.1).

Example 6.1. If

$$G(s) = \left(\frac{A|B}{C|D}\right) = \left(\begin{array}{cc|c} 0 & 1 & 0 \\ -2 & -3 & 1 \\ \hline -10 & -2 & 2 \end{array}\right), \tag{6.12}$$

then it follows from (6.10) that

$$G(s) = (-10, -2)\left(\begin{pmatrix} s & 0 \\ 0 & s \end{pmatrix} - \begin{pmatrix} 0 & 1 \\ -2 & -3 \end{pmatrix}\right)^{-1}\begin{pmatrix} 0 \\ 1 \end{pmatrix} + 2 \tag{6.13}$$

$$= (-10, -2)\begin{pmatrix} s & -1 \\ 2 & s+3 \end{pmatrix}^{-1}\begin{pmatrix} 0 \\ 1 \end{pmatrix} + 2$$

$$= \frac{1}{(s+1)(s+2)}(-10, -2)\begin{pmatrix} s+3 & 1 \\ -2 & s \end{pmatrix}^{-1}\begin{pmatrix} 0 \\ 1 \end{pmatrix} + 2$$

$$= -2\frac{s+5}{(s+1)(s+2)} + 2 = \frac{2(s+3)(s-1)}{(s+1)(s+2)}.$$

The principle above will establish the most important **cascade connection** (concatenation rule) [Kimu, (2.13), p. 15]: Given two state space representations

$$G_k(s) = \left(\frac{A_k|B_k}{C_k|D_k}\right), \quad k = 1, 2, \tag{6.14}$$

their cascade connection $G(s) = G_1(s)G_2(s)$ is given by

$$G(s) = G_1(s)G_2(s) = \left(\begin{array}{cc|c} A_1 & B_1C_2 & B_1D_2 \\ O & A_2 & B_2 \\ \hline C_1 & D_1C_2 & D_1D_2 \end{array}\right). \tag{6.15}$$

Proof of (6.15). We have the input/output relation (6.10)

$$Y(s) = G_1(s)U(s), \quad U(s) = G_2(s)V(s) \tag{6.16}$$

which means that

$$\begin{cases} \dot{\mathbf{x}} = A_1\mathbf{x} + B_1\mathbf{u}, \\ \mathbf{y} = C_1\mathbf{x} + D_1\mathbf{u} \end{cases} \tag{6.17}$$

and

$$\begin{cases} \dot{\boldsymbol{\xi}} = A_2\boldsymbol{\xi} + B_2\mathbf{v}, \\ \mathbf{u} = C_2\boldsymbol{\xi} + D_2\mathbf{v}. \end{cases} \tag{6.18}$$

Eliminating **u**, we conclude that

$$\begin{cases} \dot{\mathbf{x}} = A_1\mathbf{x} + B_1 C_2 \boldsymbol{\xi} + B_1 D_2 \mathbf{v}, \\ \mathbf{y} = C_1\mathbf{x} + D_1 C_2 \boldsymbol{\xi} + D_1 D_2 \mathbf{v}. \end{cases} \tag{6.19}$$

Hence

$$\begin{cases} \begin{pmatrix} \dot{\mathbf{x}} \\ \dot{\boldsymbol{\xi}} \end{pmatrix} = \begin{pmatrix} A_1 & B_1 C_2 \\ O & A_2 \end{pmatrix} \begin{pmatrix} \mathbf{x} \\ \boldsymbol{\xi} \end{pmatrix} + \begin{pmatrix} B_1 D_2 \\ B_2 \end{pmatrix} \mathbf{v}, \\ \mathbf{y} = \begin{pmatrix} C_1 & D_1 C_2 \end{pmatrix} \begin{pmatrix} \mathbf{x} \\ \boldsymbol{\xi} \end{pmatrix} + D_1 D_2 \mathbf{v} \end{cases} \tag{6.20}$$

whence we conclude (6.15).

Example 6.2. Given two state space representations (6.14), their parallel connection $G(s) = G_1(s) + G_2(s)$ is given by

$$G(s) = G_1(s) + G_2(s) = \left(\begin{array}{cc|c} A_1 & O & B_1 \\ O & A_2 & B_2 \\ \hline C_1 & C_2 & D_1 + D_2 \end{array} \right). \tag{6.21}$$

Indeed, we have (6.17) and for (6.18), we have

$$\begin{cases} \dot{\boldsymbol{\xi}} = A_2\boldsymbol{\xi} + B_2\mathbf{u}, \\ \mathbf{y} + \mathbf{z} = C_2\boldsymbol{\xi} + D_2\mathbf{u}. \end{cases} \tag{6.22}$$

Hence for (6.20), we have

$$\begin{cases} (\mathbf{x} + \boldsymbol{\xi})^{\cdot} = A_1\mathbf{x} + A_2\boldsymbol{\xi} + (B_1 + B_2)\mathbf{u}, \\ \mathbf{y} = C_1\mathbf{x} + C_2\boldsymbol{\xi} + (D_1 + D_2)\mathbf{v}, \end{cases} \tag{6.23}$$

whence (6.21) follows.

As an example. combining (6.15) and (6.21) we deduce

$$I - G_1(s)G_2(s) = \left(\begin{array}{ccc|c} I & O & & O \\ O & -A_1 & -B_1 C_2 & -B_1 D_2 \\ & O & -A_2 & -B_2 \\ \hline O & -C_1 & -D_1 C_2 & V^{-1} \end{array} \right). \tag{6.24}$$

Example 6.3. For (6.1), we consider the inversion $U(s) = G^{-1}(s)Y(s)$. Solving the second equality in (6.1) for **u** we obtain

$$\mathbf{u} = -D^{-1}C\mathbf{x} + D^{-1}\mathbf{y}.$$

Substituting this in the first equality in (6.1), we obtain

$$\dot{\mathbf{x}} = (A - BD^{-1}C)\mathbf{x} + BD^{-1}\mathbf{y},$$

whence

$$G^{-1}(s) = \left(\begin{array}{c|c} A - BD^{-1}C & -BD^{-1} \\ \hline -D^{-1}C & D^{-1} \end{array}\right). \tag{6.25}$$

Example 6.4. If the transfer function

$$\Theta(s) = \begin{pmatrix} \Theta_{11} & \Theta_{12} \\ \Theta_{21} & \Theta_{22} \end{pmatrix} \tag{6.26}$$

has a state space representation

$$\Theta(s) = \left(\begin{array}{c|c} A & B \\ \hline C & D \end{array}\right) = \left(\begin{array}{c|cc} A & B_1 & B_2 \\ \hline C_1 & D_{11} & D_{12} \\ C_2 & D_{21} & D_{22} \end{array}\right), \tag{6.27}$$

then we are to embed it in the linear system

$$\begin{cases} \dot{\mathbf{x}} = A\mathbf{x} + \begin{pmatrix} B_1 & B_2 \end{pmatrix} \begin{pmatrix} \mathbf{b}_1 \\ \mathbf{b}_2 \end{pmatrix}, \\ \begin{pmatrix} \mathbf{a}_1 \\ \mathbf{a}_2 \end{pmatrix} = \mathbf{y} = \begin{pmatrix} C_1 \\ C_2 \end{pmatrix} \mathbf{x} + \begin{pmatrix} D_{11} & D_{12} \\ D_{21} & D_{22} \end{pmatrix} \begin{pmatrix} \mathbf{b}_1 \\ \mathbf{b}_2 \end{pmatrix}. \end{cases} \tag{6.28}$$

6.3 Chain scattering representation

Following [Kimu, p. 7, p. 67], we first give the definition of a chain scattering representation of a system.

Suppose $\mathbf{a}_1 \in \mathbb{R}^m$, $\mathbf{a}_2 \in \mathbb{R}^q$, $\mathbf{b}_1 \in \mathbb{R}^r$ and $\mathbf{b}_2 \in \mathbb{R}^p$ are related by

$$\begin{pmatrix} \mathbf{a}_1 \\ \mathbf{a}_2 \end{pmatrix} = P \begin{pmatrix} \mathbf{b}_1 \\ \mathbf{b}_2 \end{pmatrix}, \tag{6.29}$$

where

$$P = \begin{pmatrix} P_{11} & P_{12} \\ P_{21} & P_{22} \end{pmatrix}. \tag{6.30}$$

According to the embedding principle, this is to be thought of as $\mathbf{y} = S\mathbf{u}$ corresponding to the second equality in (6.1).

(6.29) means that

$$\mathbf{a}_1 = P_{11}\mathbf{b}_1 + P_{12}\mathbf{b}_2, \quad \mathbf{a}_2 = P_{21}\mathbf{b}_1 + P_{22}\mathbf{b}_2. \tag{6.31}$$

Assume that P_{21} is a *(square) regular* matrix (whence $q = r$). Then from the second equality of (6.31), we obtain

$$\mathbf{b}_1 = P_{21}^{-1}(\mathbf{a}_2 - P_{22}\mathbf{b}_2) = -P_{21}^{-1}P_{22}\mathbf{b}_2 + P_{21}^{-1}\mathbf{a}_2. \tag{6.32}$$

Substituting (6.32) in the first equality of (6.31), we deduce that

$$\mathbf{a}_1 = \left(P_{12} - P_{11}P_{21}^{-1}P_{22}\right)\mathbf{b}_2 + P_{11}P_{21}^{-1}\mathbf{a}_2. \tag{6.33}$$

Hence putting

$$\Theta = CHAIN(P) = \begin{pmatrix} P_{12} - P_{11}P_{21}^{-1}P_{22} & P_{11}P_{21}^{-1} \\ -P_{21}^{-1}P_{22} & P_{21}^{-1} \end{pmatrix} \tag{6.34}$$

$$= \begin{pmatrix} \Theta_{11} & \Theta_{12} \\ \Theta_{21} & \Theta_{22} \end{pmatrix},$$

which is usually referred to as a **chain scattering representation** of P, we obtain an equivalent form of (6.29)

$$\begin{pmatrix} \mathbf{a}_1 \\ \mathbf{b}_1 \end{pmatrix} = CHAIN(P)\begin{pmatrix} \mathbf{b}_2 \\ \mathbf{a}_2 \end{pmatrix} = \begin{pmatrix} \Theta_{11} & \Theta_{12} \\ \Theta_{21} & \Theta_{22} \end{pmatrix}\begin{pmatrix} \mathbf{a}_2 \\ \mathbf{b}_2 \end{pmatrix}. \tag{6.35}$$

Suppose that $\mathbf{a}_2$ is fed back to $\mathbf{b}_2$ by

$$\mathbf{b}_2 = S\mathbf{a}_2, \tag{6.36}$$

where S is a **controller**. Multiplying the second equality in (6.31) by S and incorporating (6.36), we find that

$$\mathbf{b}_2 = S\mathbf{a}_2 = SP_{21}\mathbf{b}_1 + SP_{22}\mathbf{b}_2,$$

whence $\mathbf{b}_2 = (I - P_{22}S)^{-1}SP_{21}\mathbf{b}_1$.

Let the **closed-loop transfer function** Φ be defined by

$$\mathbf{a}_1 = \Phi\mathbf{b}_1. \tag{6.37}$$

Φ is given by

$$\Phi = P_{11} + P_{12}(E - P_{22}S)^{-1}SP_{21}. \tag{6.38}$$

(6.38) is sometimes referred to as a **linear fractional transformation** and denoted by

$$LF(P; S).$$

Substituting (6.36), (6.35) becomes

$$\begin{pmatrix} \mathbf{a}_1 \\ \mathbf{b}_1 \end{pmatrix} = \begin{pmatrix} \Theta_{11}S + \Theta_{12} \\ \Theta_{21}S + \Theta_{22} \end{pmatrix}\mathbf{a}_2,$$

whence we deduce that

$$\Phi = (\Theta_{11}S + \Theta_{12})(\Theta_{21}S + \Theta_{22})^{-1} = \Theta S, \qquad (6.39)$$

the **linear fractional transformation** (which is referred to as a **homographic transformation** and denoted by $HM(\Phi; S)$), where in the last equality we mean the action of Θ on the variable S. We must impose the non-constant condition $|\Theta| \neq 0$. Then $\Theta \in GL_{m+r}(\mathbb{R})$.

If S is obtained from S' under the action of Θ', $S = \Theta'S'$, then its composition Θ'' with (6.39) yields $\Theta''S' = \Phi\Phi' = \Theta\Theta'S'$, i.e.

$$\Theta'' = \Theta\Theta', \quad HM(\Theta; HM(\Theta'; S)) = HM(\Theta\Theta'; S), \qquad (6.40)$$

which is referred to as the **cascade connection** or the cascade structure of Θ and Θ'.

Thus the chain-scattering representation of a system allows us to treat the feedback connection as a cascade connection.

Suppose a closed-loop system is given with $\mathbf{z} = \mathbf{a}_1 \in \mathbb{R}^m$, $\mathbf{y} = \mathbf{a}_2 \in \mathbb{R}^q$, $\mathbf{w} = \mathbf{b}_1 \in \mathbb{R}^r$ and $\mathbf{u} = \mathbf{b}_2 \in \mathbb{R}^p$ and Φ given by (6.30).

H^∞-control problem

Find a controller S such that the closed-loop system is internally stable and the transfer function Φ satisfies

$$||\Phi||_\infty < \gamma \qquad (6.41)$$

for a positive constant γ. For the meaning of the norm, cf. §6.5.

6.4   Siegel upper space

Let $*$ denote the conjugate transpose of a square matrix: $S^* = {}^t\bar{S}$ and let the imaginary part of S defined by $\operatorname{Im} S = \frac{1}{2j}(S - S^*)$. Let $\mathcal{H}_n$ be the **Siegel upper half-space** consisting of all the matrices S (recall Eq. (6.36)) whose imaginary parts are positive definite ($\operatorname{Im} S > 0$—imaginary parts of all eigen values are positive) and satisfies $S = {}^tS$:

$$\mathcal{H}_n = \{S \in \mathrm{M}_n(\mathbb{C}) \,|\, \operatorname{Im} S > 0, \ S = {}^tS\} \qquad (6.42)$$

and let $\mathrm{Sp}(n, \mathbb{R})$ denote the **symplectic group** of order n:

$$\mathrm{Sp}(n, \mathbb{R}) \qquad (6.43)$$

$$= \left\{ \Theta = \begin{pmatrix} \Theta_{11} & \Theta_{12} \\ \Theta_{21} & \Theta_{22} \end{pmatrix} \,\middle|\, \begin{pmatrix} \Theta_{11} & \Theta_{12} \\ \Theta_{21} & \Theta_{22} \end{pmatrix}^{-1} = \begin{pmatrix} \Theta_{22} & -{}^t\Theta_{12} \\ -{}^t\Theta_{21} & \Theta_{11} \end{pmatrix} \right\}.$$

The **action** of $\mathrm{Sp}(n, \mathbb{R})$ on $\mathcal{H}_n$ is defined by (6.39) which we restate as

$$\Theta S = (\Theta_{11} S + \Theta_{12}) (\Theta_{21} S + \Theta_{22})^{-1} (= \Phi). \tag{6.44}$$

Theorem 6.1. *For a controller S living in the Siegel upper space, its rotation $Z = -jS$ lies in the right half-space $\mathcal{RHS}$. i.e. stable having positive real parts. For the controller Z, the feedback connection*

$$-j\mathbf{b}_2 = Z(-j\mathbf{a}_2) \tag{6.45}$$

is accommodated in the cascade connection of the chain scattering representation Θ (6.40), which is then viewed as the action (6.40) of $\Theta \in \mathrm{Sp}(n, \mathbb{R})$ on $S \in \mathcal{H}_n$:

$$(\Theta\Theta')S = \Theta(\Theta'S); \quad or \quad HM(\Theta; HM(\Theta'; S)) = HM(\Theta\Theta'; S), \tag{6.46}$$

where Θ is subject to the condition

$${}^t\bar{\Theta}U\Theta = U, \tag{6.47}$$

with $U = \begin{pmatrix} O & I_n \\ -I_n & O \end{pmatrix}$, An FOPID controller (in §6.6), being a unity feedback connection, is also accommodated in this framework.

Remark 6.1. With action, we may introduce the orbit decomposition of $\mathcal{H}_n$ and whence the fundamental domain. We note that in the special case of $n = 1$, we have $\mathcal{H}_1 = \mathcal{H}$ and $\mathrm{Sp}(1, \mathbb{R}) = \mathrm{SL}_n(\mathbb{R})$ and the theory of modular forms of one variable is well-known. Siegel modular forms are generalizations of the one variable case into several variables. This corresponds to the relation between SISO (single input, single output) and MIMO (multiple input, multiple output) systems. As in the case of the sushmna principle in [Vist], there is a need to rotate the upper half-space into the right half-space $\mathcal{RHS}$, which is a counter part of the right-half plane $\mathcal{RHP}$. In the case of Siegel modular forms, the matrices are constant, while in control theory, they are analytic functions (mostly rational functions analytic in $\mathcal{RHP}$). Cf. the passage in [Kimu, p. 103, ll. 2-3]. The most essential and restrictive assumption in §6.3 is that P_{21} is a *regular* matrix and it is stated [Kimu, p. 103] that Xin and Kimura succeeded in treating the case of non-singular matrices. A general theory would be useful for control theory. See §6.7 for physically realizable cases. There are many research problems lying in this direction.

The following table shows the correspondence between the control system and the Riemann zeta-function ($\zeta(s)$, $s = \sigma + j\omega$). If a shift of 1 is made, then the region of convergence coincide. In the case of the Riemann zeta-function $\zeta(s)$, the critical line used to be $\sigma = 1$ and non-vanishing on $\sigma = 1$ is equivalent to the prime number theorem (PNT). For the refined form of the PNT, the critical line is $\sigma = \frac{1}{2}$. Cf. Proposition 6.1 below.

Table 1. Correspondence between control systems and zeta-functions.

system	functions	action	region of convergence	critical line
S	rational	symplectic	$\sigma > 0$	$\sigma = 1$
s or τ	meromorphic	modular	$\sigma > 1$	$\sigma = \frac{1}{2}$

6.5 Norm of the function spaces

The norm $\mathbf{x} = \begin{pmatrix} x_1 \\ \vdots \\ x_n \end{pmatrix} \in \mathbb{C}^n$ is defined to be the Euclidean norm

$$||\mathbf{x}|| = ||\mathbf{x}||_2 = \sqrt{\sum_{j=1}^{n} |x_j|^2} \tag{6.48}$$

or by the **sup norm**

$$||\mathbf{x}|| = ||\mathbf{x}||_\infty = \max\{|x_1|, \cdots, |x_n|\}, \tag{6.49}$$

or anything that satisfies the axioms of the norm. They introduce the same topology on $\mathbb{C}^n$.

The sup norm is a limit of the p-norm as $p \to \infty$: For $\mathbf{a} = (a_1, \cdots, a_n)$,

$$\lim_{p\to\infty} ||\mathbf{a}||_p = \lim_{p\to\infty} \left(\sum_{k=1}^{n} |a_k|^p \right)^{\frac{1}{p}} = ||\mathbf{a}||_\infty = \max_{1\le k\le n} \{|a_k|^p\}. \tag{6.50}$$

For suppose $|a_1| = \max_{1\le k\le n}\{|a_k|^p\}$. Then for any $p > 0$ $|a_1| = (|a_1|^p)^{\frac{1}{p}} \le (\sum_{k=1}^{n} |a_k|^p)^{\frac{1}{p}}$.

On the other hand, since $|a_1| \ge |a_k|$, $1 \le k \le n$, we obtain

$$\left(\sum_{k=1}^{n} |a_k|^p \right)^{\frac{1}{p}} = |a_1| \left(1 + \sum_{k=2}^{n} \left| \frac{a_k}{a_1} \right|^p \right)^{\frac{1}{p}} \le |a_1|(1 + n - 1)^{\frac{1}{p}}. \tag{6.51}$$

For $p > 1$, the Bernoulli inequality gives $(1 + n - 1)^{\frac{1}{p}} \le 1 + \frac{n-1}{p} \to 1$ as $p \to \infty$. Hence the right-hand side of (6.51) tends to $|a_1|$.

The proof of (6.50) can be readily generalized to give

$$\lim_{p \to \infty} ||f||_p = ||f||_\infty = \sup_{t \geq 0} |f(t)|. \tag{6.52}$$

The p-norm in (6.52) is defined by

$$||f||_p = \left(\int_0^\infty ||f(t)||^p \, \mathrm{d}t \right)^{\frac{1}{p}},$$

where $||f(t)||$ is any Euclidean norm. Note that the functions are not ordinary functions but classes of functions which are regarded as the same if they differ only at measure 0 set. L^p is a Banach space (i.e. a complete metric space) and in particular L^2 is a Hilbert space. The 2-norm $||\cdot||_2$ is induced from the inner product

$$< f, g > = \int_0^\infty f^*(t)g(t) \, \mathrm{d}t, \quad ||f||_2 = \sqrt{< f, f >}, \tag{6.53}$$

where $*$ refers to the transposed complex conjugation.

The Parseval identity holds true if and only if the system be complete.

However, the restriction that $||f(t)|| \to 0$ as $t \to \infty$ excludes signals of infinite duration such as unit step signals or periodic ones from L_p. To circumvent the inconvenience, the notion of averaged norm, or the mean square value

$$M_2(f) = M_2(f, T) = \frac{1}{T} \int_0^T ||f(t)||^2 \, \mathrm{d}t \tag{6.54}$$

or similar, has been introduced and intensively studied both in analytic number theory and control theory, in the latter of which the **power** norm has been introduced:

$$\text{power}(f) = \lim_{T \to \infty} M_2(f, T)^{\frac{1}{2}} = \lim_{T \to \infty} \left(\frac{1}{T} \int_0^T ||f(t)||^2 \, \mathrm{d}t \right)^{\frac{1}{2}}. \tag{6.55}$$

Remark 6.2. In mathematics and in particular in analytic number theory, studying the mean square or higher power moments in the form of a sum or an integral is quite common. Especially, this idea is applied to finding out the true order of magnitude of the error term on average. Such an average result will give a hint on the order of the error term itself.

Let $\zeta(s)$ denote the Riemann zeta-function having a simple pole at $s = 1$. It is essential that it does not vanish on the line $\sigma = 1$ for the prime number theorem (PNT; cf. (3.72)) to hold. The plausible best bound for the error term for the PNT is equivalent to the celebrated **Riemann hypothesis**

(RH) to the effect that the Riemann zeta-function does not vanish on the critical line $\sigma = \frac{1}{2}$. Since the values on the critical line are expected to be small, the averaged norm $M_2(\zeta)$ or $M_4(\zeta)$, i.e. the mean value

$$M_{2k}(\zeta) = \frac{1}{T} \int_0^T \left| \zeta\left(\frac{1}{2} + it\right) \right|^{2k} dt \tag{6.56}$$

for $k \geq 1$ is of great interest and there have appeared a great deal of research on the subject. The first result for $M_4(\zeta)$ is due to Ingham who used the approximate functional equation for the Riemann zeta-function to obtain (6.57).

Proposition 6.1.

$$M_4(\zeta) = \frac{1}{T} \int_0^T \left| \zeta\left(\frac{1}{2} + it\right) \right|^4 dt = \frac{1}{4\pi^2} \log^4 T(1 + o(1)) \tag{6.57}$$

for $T \to \infty$. See e.g. [Bell]. The main interest in such estimates as (6.57) lies in the fact that estimates for all $k \in \mathbb{N}$

$$M_{2k}(\zeta) = O(T^a) \tag{6.58}$$

would imply the weak Lindelöf hypothesis (LH) in the form

$$\zeta\left(\frac{1}{2} + it\right) = O(T^{\frac{a}{2k} + \varepsilon}) \tag{6.59}$$

for every $\varepsilon > 0$. It is apparent that the RH implies the LH.

Cf. [Ivic] and [Titc] for more details on power moments.

The **Hardy space** H^p (cf. e.g. [Kimu, p. 39]) is well-known. It consists of all $f(s)$ which are analytic in $\mathcal{RHP}$—right half-plane $\sigma > 0$ such that $f(j\omega) \in L^p$; in particular, H^∞ with sup norm. Thus H^∞-control problem is about those (rational) functions which are analytic in $\mathcal{RHP}$, *a fortiori* stable, with regard to the sup norm. Thus the above-mentioned mean-value problem for the Riemann zeta-function corresponds to the H^{2k}-control problem with transfer functions which are mostly rational functions. Thus we might say that the $2k$-th mean values of the finite Dirichlet series (main ingredients in the approximate functional equation) corresponds to the H^{2k}-control problem. As is known, if all the mean values are $O(T^\varepsilon)$, then the (strong) Lindelöf hypothesis would follow. Thus, all these are centered around the LH. Asking for (almost) all individual values, the H^∞-control problem corresponds to the weak LH and eventually to the RH.

Remark 6.3. The H^∞-control is often referred to as *robust* suggesting that the error estimate is invulnerable because it estimates all individual values rather than average values in the case of H^{2k}. However, what prevails here is finitely many sample values taken at a certain time interval which are then approximated by the pre-assigned finitely many scales, thus making everything *quantized*. In practice therefore, what is needed is *craftsman's intuition* to choose right time intervals and right scales, and above all things, the proper transfer function. It would be desirable if we can find a means to find a proper class of special functions whose rational approximation will give a proper transfer function.

6.6   (Unity) feedback system

The synthesis problem of a controller of the **unity feedback system,** depicted in figure below, refers to the **sensitivity reduction problem,**

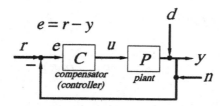

which asks for the estimation of the **sensitivity function** $S = S(s)$ multiplied by an appropriate frequency weighting function $W = W(s)$:

$$S = (I + PC)^{-1}, \tag{6.60}$$

is a transfer function from r to e, where $C = K$ is a **compensator** and P is a plant. The problem consists in reducing the magnitude of S over a specified frequency range Ω, which amounts to finding a compensator C stabilizing the closed-loop system such that

$$\|WS\|_\infty < \gamma \tag{6.61}$$

for a positive constant γ.

To accommodate this in the H^∞ control problem (6.29), we choose the matrix elements P_{ij} of P in such a way that the closed-loop transfer function Φ in (6.38) coincides with WS. First we are to choose $P_{22} = -P$. Then we would choose $P_{12}P_{21} = WP$. Then Φ becomes $P_{11} + WPC(I + PC)^{-1} = P_{11} - W + W(I + PC)^{-1}$. Hence choosing $P_{11} = W$, we have $\Phi = WS$.

Hence we may choose e.g.

$$P = \begin{pmatrix} P_{11} & P_{12} \\ P_{21} & P_{22} \end{pmatrix} = \begin{pmatrix} W & WP \\ E & -P \end{pmatrix}. \tag{6.62}$$

Example 6.5. First we treat the case of general feedback scheme. Denoting the Laplace transforms by the corresponding capital letters, we have

$$Y = PR + PU, U = KE,$$

whence $Y = PR + PKE$. Now if it so happens that $e = r - y$ and P is replaced by PK, i.e. in the case of unity FD, we derive (6.60) directly from the figure below

outputs   **inputs**

We have $E = R - Y$, so that $Y = PR + PK(R - Y)$. Solving in Y, we deduce that $(I + PK)^{-1} PKR$.

We take into account the disturbance d, we obtain since $U = CE = C(R - Y)$

$$Y = PU + PD = PC(R - Y) + PD,$$

whence $Y = PC(R - Y) + PD$. It follows that $Y = (I + PC)^{-1} PCR + (I + PC)^{-1} PD$. In the case where $d = 0$, PC being the open loop transfer function, we have SR is the tracking error for the input R. Hence (6.60) holds true.

6.7 *J*-lossless factorization and dualization

In this section we partially follow Helton ([Hel1], [Hel2], [HeMe]) who uses the unit ball in place of $\mathcal{RHP}$. They shift to each other under the complex exponential map. For conventional control theory, the unit ball is to be replaced by the critical line ($\sigma = 0$). In practice what appears is the algebra of functions ([Hel2, p.2]).

$\mathcal{R} = \{$functions defined on the unit ball having the rational continuation

to the whole space$\}$ (6.63)

or still larger algebra ψ consisting of those functions which have (pseudo-) meromorphic continuations ([Hel2, footnote 6, p. 27]). The occurrence of the gamma function [Hel2, Fig. 2.5, p. 17] justifies our incorporation of more advanced special functions and ultimately zeta-functions in control theory (see §6.11).

Along with the algebra $\mathcal{R}$, one considers

$$\mathcal{B}H^\infty = \{F|\text{analytic on the unit ball having the supremum norm} < 1\}.$$
$$(6.64)$$

Then the only mapping $\Theta \in \mathcal{R}U(m, r)$ acting on $\mathcal{B}H^\infty$ must satisfy the J-lossless property: Let Θ denote an $(m + r) \times (m + r)$ matrix. Then

$$\Theta^* J_{mr} \Theta \le J_{mr},\qquad(6.65)$$

where $J_{mr} = \begin{pmatrix} I_m & O \\ O & -I_r \end{pmatrix}$ is a **signature matrix**, with I_m indicating the identity matrix of degree m. (6.65) is interpreted to be the power preservation of the system in the chain scattering representation (6.34) ([Kimu, p. 82]). To explain these we state

Definition 6.2. If in the system $\Sigma = \Sigma(s)$ in (6.29) is stable and the input power is always equal to the output power, i.e.

$$||\mathbf{b}_1(j\omega)||^2 + ||\mathbf{b}_2(j\omega)||^2 = ||\mathbf{a}_1(j\omega)||^2 + ||\mathbf{a}_2(j\omega)||^2,\qquad(6.66)$$

then the system is said to be **lossless**.

In terms of (6.35), (6.66) reads

$$||\mathbf{a}_1(j\omega)||^2 - ||\mathbf{b}_1(j\omega)||^2 = ||\mathbf{b}_2(j\omega)||^2 + ||\mathbf{a}_2(j\omega)||^2,$$

which in turn amounts to (6.65) with equality sign:

$$^t\Theta(-j\omega)J_{mr}\Theta(j\omega) = J_{pr}.\qquad(6.67)$$

In view of the signature matrix J, a power preserving system Σ is said to be J-lossless, or more precisely, the matrix Θ satisfying both (6.67) and

$$\Theta^*(s)J_{mr}\Theta(s) \le J_{pr},\qquad(6.68)$$

in the $\mathcal{RHP}$, is called (J_{mr}, J_{pr})-lossless or (J, J')-lossless.

Remark 6.4. As is mentioned in Abstract, it is not clear what is the counterpart of the functional equation in control theory since the transfer function is too simple to have a symmetry property. A plausible candidate for the symmetry looks like J-lossless conjugation, Cf. Siegel [Sieg].

Definition 6.3. A J-loss matrix $\Theta = \Theta(s)$ is called a **stabilizing conjugator** of a transfer function $G = G(s)$ if

(i) ΘG is stable

and

(ii) the degree of Θ is equal to the number of unstable poles of G, multiplicity being counted.

Condition (i) means that all unstable poles of G are cancelled by the zeros of Θ and (ii) means that the degree of Θ is minimal to achieve (i), i.e. the zeros of G are not cancelled by the poles of Θ. The latter situation does not match the case of the Riemann zeta-function ζ, where the zeros at negative even integers of ζ are *cancelled by* the poles of the gamma function $\Gamma\left(\frac{s}{2}\right)$. The shifting effect of Condition (i) of an unstable pole into stable one seems to have a similar effect as the reflection of with respect to the critical line; however, this is done by multiplication of a stabilizing conjugator, which has some arbitrariness. This probably means that we are to seek the *niryana* of the transfer function and the gamma factor if any whose approximation works as the stabilizing conjugator. As can be seen in [Hel2], there is associated the notion of hyperbolic geometry to control systems, as is the same with automorphic forms, there seems to lie a rich field of research in the direction.

Finally we briefly refer to the dual chain-scattering representation of the plant P in (6.30). We assume P_{12} is a square invertible matrix (whence $m = p$). Then the argument goes in parallel to that leading to (6.35). Defining the **dual chain scattering matrix** by

$$DCHAIN(P) = \begin{pmatrix} P_{12}^{-1} & P_{11}P_{12}^{-1} \\ -P_{12}^{-1}P_{22}\,P_{21} - P_{22}P_{12}^{-1}P_{11} \end{pmatrix}, \tag{6.69}$$

we obtain

$$CHAIN(P) \cdot DCHAIN(P) = E. \tag{6.70}$$

6.8 FOPID controllers

"FO" means "Fractional order" and "PID" refers to "Proportional, Integral, Differential", where "Proportional" means just constant times the input function $e(t)$, "Integral" means the **fractional order integration** $I_t^{\lambda} D_t^{-\lambda}$ of $e(t)$ $(\lambda > 0)$, and "Differential" the **fractional order differentiation** D_t^{δ} of $e(t)$ $(\delta > 0)$.

The FO $PI^\lambda D^\delta$ controller (control signal in the time domain) is one of the most refined feed-forward compensator defined by

$$u(t) = \left(K_p + K_i D_t^{-\lambda} + K_d D_t^\delta\right) e(t), \tag{6.71}$$

where u is the input function, e is the deviation and K_p, K_i, K_d are constant parameters which are to be specified (K_p- the position feedback gain, K_d- the velocity feedback gain). DE (6.71) translates into the **state equation**

$$Y(s) = C(s)E(s), \tag{6.72}$$

where U, Y indicate the Laplace transforms of u, y, respectively (cf. §6.9) and G is the compensator **continuous transfer function**

$$C(s) = K_p + K_i s^{-\lambda} + K_d s^\delta. \tag{6.73}$$

The derivation of (6.73) from (6.71) depends on the following. The general fractional calculus operator ${}_a D_t^\alpha$ is symbolically stated as

$$
{}_a D_t^\alpha = \begin{cases} \frac{\mathrm{d}^\alpha}{\mathrm{d}t^\alpha}, & \operatorname{Re}\alpha > 0 \\ 1, & \operatorname{Re}\alpha = 0 \\ \int_a^t \frac{1}{\mathrm{d}t^\alpha}, & \operatorname{Re}\alpha < 0, \end{cases} \tag{6.74}
$$

where a and t are the lower and upper limits of integration and α is the order of calculus.

More precisely, the definition of the fractional differo-integral is given by the Riemann-Liouville expression

$$
{}_a D_t^\alpha f(t) = \frac{1}{\Gamma(1 - \{\alpha\})} \left(\frac{\mathrm{d}}{\mathrm{d}t}\right)^{\alpha - \{\alpha\} + 1} \int_a^t (t - \tau)^{-\{\alpha\}} f(\tau)\, \mathrm{d}\tau \tag{6.75}
$$

where $\{\alpha\} = \alpha - [\alpha]$ indicates the fractional part of α, with $[\alpha]$ the integral part of α. Thus we are also led to the **Riemann-Liouville fractional integral transform:**

$$
\mathcal{RL}[f] = \frac{1}{\Gamma(\mu)} \int_0^y (y - x)^{\mu - 1} f(x)\, \mathrm{d}x. \tag{6.76}
$$

For applications, cf. §6.11.

When $\alpha \in \mathbb{N}$, (6.75) reads

$$
{}_a D_t^\alpha f(t) = \left(\frac{\mathrm{d}}{\mathrm{d}t}\right)^{\alpha + 1} \int_a^t f(\tau)\, \mathrm{d}\tau = f^{(\alpha)}(t), \tag{6.77}
$$

the α-th derivative of f.

We shall see that the definition (6.75) is a natural outcome of the general formula for the difference operator of order $\alpha \in \mathbb{N}$ with difference $y \geq 0$:

$$\Delta_y^\alpha f(x) = \sum_{\nu=0}^{\alpha} (-1)^{\alpha-\nu} \binom{\alpha}{\nu} f(x + \nu y). \tag{6.78}$$

If f has the α-th derivative $f^{(\alpha)}$, then

$$\Delta_y^\alpha f(x) = \int_x^{x+y} dt_1 \int_{t_1}^{t_1+y} dt_2 \cdots \int_{t_{\alpha-1}}^{t_{\alpha-1}+y} f^{(\alpha)}(t_\alpha)\, dt_\alpha. \tag{6.79}$$

The special case of (6.79) with $t_\nu = a, a + y \to x$ ($\varphi(t) = f^{(\alpha)}(t)$) reads

$$\Delta_{x-a}^\alpha \varphi(x) = \int_a^x dt \int_a^x dt \cdots \int_a^x \varphi(t)\, dt = \frac{1}{\Gamma(\alpha)} \int_a^x (x-t)^{\alpha-1} \varphi(t)\, dt, \tag{6.80}$$

whose far right-hand side is $\mathcal{R}L[\varphi]$.

Let $F(s)$ be the Laplace transform of the input function $f(t)$. Then

$$L[_0D_t^\alpha f](t) = s^\alpha F(s) - {}_0D_t^{\alpha-1} f(t)\,|_{t=0} \tag{6.81}$$

and

$$L[_0D_t^{-\alpha} f](t) = s^{-\alpha} F(s). \tag{6.82}$$

Both of H^∞- and PID-controllers are applied successfully in the EV control by Cao [CaoC], [YBCa], which we may unify in our framework.

6.9 Fourier, Mellin and (two-sided) Laplace transforms

We state the Mellin, (two-sided) Laplace and the Fourier transforms. If $f(x) = O(x^\alpha)$, $\alpha \in \mathbb{R}$ for $x > 0$, then its **Mellin transform** $M[f]$ is defined by

$$M[f](s) = \int_0^\infty x^s f(x) \frac{dx}{x}, \quad \sigma > \alpha. \tag{6.83}$$

Under the change of variable $x = e^{-t}$, the Mellin transform and the **two-sided Laplace transform** shift each other:

$$L^\pm[\varphi](s) = \int_{-\infty}^\infty e^{-st} \varphi(t)\, dt, \quad \sigma > \alpha, \tag{6.84}$$

where we write $\varphi(t) = f(e^{-t})$.

The ordinary **Laplace transform** (one-sided Laplace transform) is obtained by multiplying the integrand by the unit step function $u = u(t)$ (cf. the passage immediately after (6.7)):

$$L[f](s) = L^{\pm}[fu](s) = \int_0^{\infty} e^{-st} f(e^{-t})\, dt, \quad \sigma > \alpha. \tag{6.85}$$

Cf. Definition 6.4 below.

If we fix $\varkappa > \alpha$ and write $s = \varkappa + j\omega$, $G(y) = L^{\pm}[f](\varkappa + j\omega)$, $g(t) = e^{-\varkappa t} f(e^{-t})$ in (6.84), then it changes into

$$G(\omega) = F[g](\omega) = \int_{-\infty}^{\infty} e^{-j\omega t} g(t)\, dt = L^{\pm}[\varphi](j\omega), \tag{6.86}$$

the **Fourier transform** of g.

We explain Plancherel's theorem for functions in $L_2(\mathbb{R})$. Let

$$\hat{f}_T(x) = \frac{1}{\sqrt{2\pi}} \int_{-T}^{T} e^{-ixt} f(t)\, dt. \tag{6.87}$$

Then $\hat{f}_T(x)$ is convergent to a function $\hat{f}$ in L^2:

$$\|\hat{f}_T - \hat{f}\| \to 0 \ \ T \to \infty; \quad \text{l.i.m.}_{T \to \infty} \hat{f}_T(x) = \hat{f}(t), \tag{6.88}$$

where l.i.m. is a short-hand for "limit in the mean". The **Parseval identity** reads

$$\|\hat{f}\|_2 = \|f\|_2, \quad \int_{-\infty}^{\infty} |\hat{f}(t)|^2\, dt = \int_{-\infty}^{\infty} |f(t)|^2\, dt. \tag{6.89}$$

If we apply (6.89) to a causal function f, then it leads to [Kimu, (3.19)]

$$\int_{-\infty}^{\infty} |L^{\pm}[f](i\omega)|^2\, d\omega = \int_0^{\infty} |f(t)|^2\, dt. \tag{6.90}$$

Hence we see that [Kimu, (3.19)] is indeed the Parseval identity for the Fourier (or Plancherel) transform for $f \in L_2(\mathbb{R})$.

6.10 Examples of second-order systems and their solution

- Electrical circuits
  The electric current $i = i(t)$ flowing an electrical circuit which consists of four ingredients, electromotive-force $e = e(t)$, resistance R, coil L and condenser C satisfies

$$L\frac{d^2 i}{dt^2} + R\frac{di}{dt} + \frac{1}{C}i = e'(t). \tag{6.91}$$

- Newton's equation of motion (cf. [Grod])

$$M\frac{d^2y}{dt^2} + R\frac{dy}{dt} + Ky = e(t) = F, \qquad (6.92)$$

where M is the inertance of mass, R is the viscous resistance of the dashpot and K is the spring stiffness.

Introducing the new parameters

$$\omega_n = \sqrt{\frac{K}{M}} : \text{natural angular frequence}$$

$$\zeta = \frac{R}{\sqrt{2\frac{K}{M}}} : \text{damping ratio,}$$

(6.92) becomes

$$\frac{1}{\omega_n^2}\frac{d^2y}{dt^2} + \frac{2\zeta}{\omega_n}\frac{dy}{dt} + y = \frac{1}{K}F. \qquad (6.93)$$

To solve (6.91), we use the Laplace transform which has been defined by (6.85) and we state its definition independently.

Definition 6.4. Suppose $y(t) = O(e^{at})$, $t \to \infty$ for an $a \in \mathbb{R}$. The Laplace transform $Y(s) = L[y](s)$ of $y = y(t)$ is defined by

$$L[y](s) = \int_0^\infty e^{-st}y(t)\,dt, \quad \operatorname{Re} s > a. \qquad (6.94)$$

The integral converges absolutely in $\operatorname{Re} s > a$ and represents an analytic function there.

It is customary to denote the Laplace transform of a function y by the corresponding capital letter Y and we follow this tradition in what follows.

Example 6.6. Let $\alpha \in \mathbb{C}$. Then

$$L[e^{\alpha t}](s) = \frac{1}{s - \alpha}, \qquad (6.95)$$

valid for $\operatorname{Re} s > \operatorname{Re}\alpha$ in the first instance. The right-hand side of (6.95) gives a **meromorphic continuation** of the left-hand side to the punctured domain $\mathbb{C}\backslash\{\alpha\}$. Furthermore, (6.95) with α replaced by $i\alpha$ reads

$$L[\sin\alpha t](s) = \frac{\alpha}{s^2 + \alpha^2} \qquad (6.96)$$

and

$$L[\cos \alpha t](s) = \frac{s}{s^2 + \alpha^2}. \tag{6.97}$$

For $\alpha = \omega \in \mathbb{R}$ they reduce to familiar formulas:

$$L[\sin \omega t](s) = \frac{\omega}{s^2 + \omega^2}$$

and

$$L[\cos \omega t](s) = \frac{s}{s^2 + \omega^2}.$$

Proof. By definition, (6.95) clearly holds true. Since the right-hand side is analytic in $\mathbb{C}\backslash\{\alpha\}$, the **consistency theorem** establishes the last assertion. Once (6.95) is established, we have

$$L[e^{i\alpha t}](s) = \frac{1}{s - i\alpha}, \quad L[e^{-i\alpha t}](s) = \frac{1}{s + i\alpha} \tag{6.98}$$

whence e.g.

$$L[\cos \alpha t](s) = \frac{1}{2}\left(L[e^{i\alpha t}](s) + L[e^{-i\alpha t}](s)\right) = \frac{s}{s^2 + \alpha^2}$$

by Euler's identity, i.e. (6.97).

Now we shall give examples of (6.2) for the second-order systems which do not appear anywhere else save for [Vist].

Example 6.7. The output signal (current) $y = y(t)$ described by the DE

$$y'' + y' + y = u(t) = e^{-\frac{1}{2}t} \sin \frac{\sqrt{3}}{2}t,$$

where $u(t)$ is the input function and the initial values are assumed to be 0: $y(0) = 0$, $y'(0) = 0$, is

$$\frac{2}{\sqrt{3}}y(t) = \frac{2}{\sqrt{3}}L^{-1}[L[y]](t) = -\frac{2}{3}te^{-\frac{1}{2}t}\cos \frac{\sqrt{3}}{2}t + \frac{2}{3\sqrt{3}}e^{-\frac{1}{2}t}\sin \frac{\sqrt{3}}{2}t.$$

Indeed, we have

$$Y(s) = \Phi(s)U(s), \quad U(s) = L[u](s) = \frac{\frac{\sqrt{3}}{2}}{s^2 + s + 1} \tag{6.99}$$

or

$$\frac{2}{\sqrt{3}}L[y](s) = \frac{1}{(s^2 + s + 1)^2}.$$

As a transfer function, the function in (6.99)

$$\Phi(s) = \frac{1}{s^2 + s + 1}$$

is stable.

Example 6.8. In the same vein as with Example 6.7, we find the solution

$$\frac{2}{\sqrt{3}}y(t) = \frac{2}{\sqrt{3}}L^{-1}[L[y]](t) = -te^{\frac{1}{2}t}\cos\frac{\sqrt{3}}{2}t + \sqrt{3}e^{\frac{1}{2}t}\sin\frac{\sqrt{3}}{2}t,$$

of the DE

$$y'' - y' + y = u(t) = e^{\frac{1}{2}t}\sin\frac{\sqrt{3}}{2}t,$$

where the initial values are assumed to be 0: $y(0) = 0$, $y'(0) = 0$.

We have

$$Y(s) = \Phi_1(s)U(s), \quad U(s) = L[u](s) = \frac{\frac{\sqrt{3}}{2}}{s^2 - s + 1} \tag{6.100}$$

or

$$\frac{2}{\sqrt{3}}L[y](s) = \frac{1}{(s^2 - s + 1)^2}$$

and the transfer function in (6.100)

$$\Phi_1(s) = \frac{1}{s^2 - s + 1}$$

is unstable.

6.11 The product of zeta-functions: ΓΓ-type

In this section, we illustrate the use of fractional integrals by proving a slight generalization of the result of Chandrasekharan and Narasimhan ([ChNa]) involving the ΓΓ-type functional equation, which is the first instance beyond Hecke theory of the functional equation with a single gamma factor. First we state the basic settings.

6.11.1 *Statement of the situation*

Let $\{\lambda_k\}$, $\{\mu_k\}$ be increasing sequences of positive numbers tending to ∞, and let $\{\alpha_k\}$, $\{\beta_k\}$ be complex sequences. We form the Dirichlet series

$$\varphi(s) = \sum_{k=1}^{\infty}\frac{\alpha_k}{\lambda_k^s}, \tag{6.101}$$

$$\psi(s) = \sum_{k=1}^{\infty}\frac{\beta_k}{\mu_k^s} \tag{6.102}$$

and suppose that they have finite abscissas of absolute convergence σ_φ, σ_ψ, respectively.

We suppose the existence of the meromorphic function χ satisfying the functional equation (of $\Gamma\Gamma$-type) of the form with r a real number, and having a finite number of poles s_k $(1 \le k \le L)$.

$$
\chi(s) = \begin{cases} \Gamma\left(s + \frac{\nu}{2}\right)\Gamma\left(s - \frac{\nu}{2}\right)\varphi(s), & (\operatorname{Re} s > \sigma_\varphi) \\ \Gamma\left(r - s + \frac{\nu}{2}\right)\Gamma\left(r - s - \frac{\nu}{2}\right)\psi(r - s), & (\operatorname{Re} s < r - \sigma_\psi). \end{cases}
$$
(6.103)

We introduce the processing gamma factor

$$
\Delta(w) = \frac{\Gamma\left(\{b_j + B_j w\}_{j=1}^m\right)\Gamma\left(\{a_j - A_j w\}_{j=1}^n\right)}{\Gamma\left(\{a_j + A_j w\}_{j=n+1}^p\right)\Gamma\left(\{b_j - B_j w\}_{j=m+1}^q\right)} \quad (A_j, B_j > 0),
$$
(6.104)

and suppose that for any real numbers u_1, u_2 $(u_1 < u_2)$

$$
\lim_{|v| \to \infty} \Delta(u + iv - s)\chi(u + iv) = 0,
$$
(6.105)

uniformly in $u_1 \le u \le u_2$.

In the w-plane we take two deformed Bromwich paths

$$
L_1(s) : \gamma_1 - i\infty \to \gamma_1 + i\infty, \quad L_2(s) : \gamma_2 - i\infty \to \gamma_2 + i\infty \quad (\gamma_2 < \gamma_1)
$$

such that they squeeze a compact set $\mathcal{S}$ with boundary $\mathcal{C}$ for which $s_k \in \mathcal{S}$ $(1 \le k \le L)$ and all the poles of

$$
\frac{\Gamma\left(\{b_j - B_j s + B_j w\}_{j=1}^m\right)}{\Gamma\left(\{a_j - A_j s + A_j w\}_{j=n+1}^p\right)\Gamma\left(\{b_j + B_j s - B_j w\}_{j=m+1}^q\right)}
$$

lie to the left of $L_2(s)$, and those of

$$
\frac{\Gamma\left(\{a_j + A_j s - A_j w\}_{j=1}^n\right)}{\Gamma\left(\{a_j - A_j s + A_j w\}_{j=n+1}^p\right)\Gamma\left(\{b_j + B_j s - B_j w\}_{j=m+1}^q\right)}
$$

lie to the right of $L_1(s)$.

Then we define the H-function by $(0 \le n \le p, 0 \le m \le q, A_j, B_j > 0)$

$$
H_{p,q}^{m,n}\left(z \left| \begin{matrix} (1 - a_1, A_1), \ldots, (1 - a_n, A_n), (a_{n+1}, A_{n+1}), \ldots, (a_p, A_p) \\ (b_1, B_1), \ldots, (b_m, B_m), (1 - b_{m+1}, B_{m+1}), \ldots, (1 - b_q, B_q) \end{matrix} \right. \right)
$$
(H-1)

$$
= \frac{1}{2\pi i} \int_L \frac{\Gamma(b_1 + B_1 s, \ldots, b_m + B_m s)\Gamma(a_1 - A_1 s, \ldots, a_n - A_n s)}{\Gamma(a_{n+1} + A_{n+1}s, \ldots, a_p + A_p s)\Gamma(b_{m+1} - B_{m+1}s, \ldots, b_q - B_q s)} z^{-s}\, ds.
$$

In the special case where $A_j = B_j = 1$, the H-function reduces to G-functions and denoted by G with other parameters remaining the same. We also define the χ-function $X(z, s)$ by

$$
X(z, s) = \frac{1}{2\pi i} \int_{L_1(s)} \Delta(w - s)\chi(w) z^{-w}\, dw,
$$
(6.106)

which is for $\chi = 1$ one of H-functions. Hereafter we always assume that $z > 0$, which may be extended to $\operatorname{Re} z > 0$.

Then we have

$$X(z, s) = \frac{1}{2\pi i} \int_{L_2(s)} \Delta(w - s)\chi(w)z^{-w} \, dw + \frac{1}{2\pi i} \int_{C} \Delta(w - s)\chi(w)z^{-w} \, dw,$$
(6.107)

which amounts to

Theorem 6.2. ([MRSK]) *We have the modular relation equivalent to* (6.103):

$$X(z, s) = \tag{6.108}$$

$$\begin{cases}
\sum_{k=1}^{\infty} \frac{\alpha_k}{\lambda_k^s} \\
\quad \times H_{p,q+2}^{m+2,n}\left(z\,\lambda_k \;\middle|\; \begin{matrix} \{(1 - a_j, A_j)\}_1^n, \{(a_j, A_j)\}_{n+1}^p \\ (s + \frac{\nu}{2}, 1), (s - \frac{\nu}{2}, 1), \{(b_j, B_j)\}_1^m, \{(1 - b_j, B_j)\}_{m+1}^q \end{matrix} \right) \\
\sum_{k=1}^{\infty} \frac{\beta_k}{\mu_k^{r-s}} \\
\quad \times H_{q,p+2}^{n+2,m}\left(\frac{\mu_k}{z} \;\middle|\; \begin{matrix} \{(1 - b_j, B_j)\}_1^m, \{(b_j, B_j)\}_{m+1}^q \\ (r - s + \frac{\nu}{2}, 1), (r - s - \frac{\nu}{2}, 1), \{(a_j, A_j)\}_1^n, \{(1 - a_j, A_j)\}_{n+1}^p \end{matrix} \right) \\
\quad + \sum_{k=1}^{L} \operatorname{Res}\left(\Delta(w - s)\chi(w)z^{s-w}, w = s_k \right).
\end{cases}$$

$$\left(\sum_{j=1}^{n} A_j + \sum_{j=1}^{m} B_j + 2 \geq \sum_{j=n+1}^{p} A_j + \sum_{j=m+1}^{q} B_j \right).$$

In the special case where $A_j = B_j = 1$, we have

Theorem 6.3.

$$z^s X(z, s) \tag{6.109}$$

$$= \begin{cases}
\sum_{k=1}^{\infty} \frac{\alpha_k}{\lambda_k^s} G_{p,q+2}^{m+2,n}\left(z\lambda_k \;\middle|\; \begin{matrix} 1 - a_1, \ldots, 1 - a_n, a_{n+1}, \ldots, a_p \\ s + \frac{\nu}{2}, s - \frac{\nu}{2}, b_1, \ldots, b_m, 1 - b_{m+1}, \ldots, 1 - b_q \end{matrix} \right) \\
\sum_{k=1}^{\infty} \frac{\beta_k}{\mu_k^{r-s}} \\
\quad \times G_{q,p+2}^{n+2,m}\left(\frac{\mu_k}{z} \;\middle|\; \begin{matrix} 1 - b_1, \ldots, 1 - b_m, b_{m+1}, \ldots, b_q \\ r - s + \frac{\nu}{2}, r - s - \frac{\nu}{2}, a_1, \ldots, a_n, 1 - a_{n+1}, \ldots, 1 - a_p \end{matrix} \right) \\
\quad + \sum_{k=1}^{L} \operatorname{Res}\left(\Delta(w - s)\chi(w) \, z^{s-w}, w = s_k \right)
\end{cases}$$

$$(2n + 2m + 2 \geq p + q).$$

For many important applications, cf. [MRSK].

6.11.2   *The Riesz sum: $G_{4,4}^{2,2} \leftrightarrow G_{2,6}^{4,0}$*

Formula (6.109) in the special case of the title reads

$$\sum_{k=1}^{\infty} \frac{\alpha_k}{\lambda_k^s} G_{4,4}^{2,2}\left(z\lambda_k \,\middle|\, \begin{matrix} a,b,c,d \\ s+\frac{\nu}{2}, s-\frac{\nu}{2}, e, f \end{matrix}\right) \tag{6.110}$$

$$= \sum_{k=1}^{\infty} \frac{\beta_k}{\mu^{r-s}} G_{2,6}^{4,0}\left(\frac{\mu_k}{z} \,\middle|\, \begin{matrix} 1-e, 1-f \\ r-s+\frac{\nu}{2}, r-s-\frac{\nu}{2}, 1-a, 1-b, 1-c, 1-d \end{matrix}\right)$$

$$+ \sum_{k=1}^{L} \text{Res}\left(\Delta(w-s)\chi(w)z^{s-w}, w=s_k\right),$$

where

$$\Delta(w) = \frac{\Gamma(1-a-w)\Gamma(1-b-w)}{\Gamma(c+w)\Gamma(d+w)\Gamma(1-e-w)\Gamma(1-f-w)}. \tag{6.111}$$

We treat the case $r = \frac{1}{2}$. Assuming λ is a non-negative *integer*, we put $a = s + \frac{\nu}{2} + \frac{\lambda}{2} + \frac{1}{2}, b = s + \frac{\nu}{2} + \frac{\lambda}{2} + 1, c = s - \frac{\nu}{2}, d = s + \frac{\nu}{2} + \lambda + 1, e = s + \frac{\nu}{2} + \frac{1}{2}, f = s + \frac{\nu}{2} + \lambda + 1$. Then (6.110) becomes

$$\sum_{k=1}^{\infty} \frac{\alpha_k}{\lambda_k^s} G_{4,4}^{2,2}\left(z\lambda_k \,\middle|\, \begin{matrix} s+\frac{\nu}{2}+\frac{\lambda}{2}+\frac{1}{2}, s+\frac{\nu}{2}+\frac{\lambda}{2}+1, s-\frac{\nu}{2}, s+\frac{\nu}{2}+\lambda+1 \\ s+\frac{\nu}{2}, s-\frac{\nu}{2}, s+\frac{\nu}{2}+\frac{1}{2}, s+\frac{\nu}{2}+\lambda+1 \end{matrix}\right)$$

$$\tag{6.112}$$

$$= \sum_{k=1}^{\infty} \frac{\beta_k}{\mu^{\frac{1}{2}-s}} G_{2,6}^{4,0}\left(\frac{\mu_k}{z} \,\middle|\, \begin{matrix} -s-\frac{\nu}{2}+\frac{1}{2}, -s-\frac{\nu}{2}-\lambda \\ * \end{matrix}\right)$$

$$+ \sum_{k=1}^{L} \text{Res}\left(\Delta(w-s)\chi(w)z^{s-w}, w=s_k\right),$$

where $*$ indicates $-s+\frac{\nu}{2}+\frac{1}{2}, -s-\frac{\nu}{2}+\frac{1}{2}, -s-\frac{\nu}{2}-\frac{\lambda}{2}+\frac{1}{2}, -s-\frac{\nu}{2}-\frac{\lambda}{2}, -s+\frac{\nu}{2}+1, -s-\frac{\nu}{2}-\lambda$.

We note that the G-functions in (6.112) reduce to

$$G_{4,4}^{2,2}\left(z \,\middle|\, \begin{matrix} s+\frac{\nu}{2}+\frac{\lambda}{2}+\frac{1}{2}, s+\frac{\nu}{2}+\frac{\lambda}{2}+1, s-\frac{\nu}{2}, s+\frac{\nu}{2}+\lambda+1 \\ s+\frac{\nu}{2}, s-\frac{\nu}{2}, s+\frac{\nu}{2}+\frac{1}{2}, s+\frac{\nu}{2}+\lambda+1 \end{matrix}\right)$$

$$= 2^\lambda G_{1,1}^{1,0}\left(\sqrt{z} \,\middle|\, \begin{matrix} 2s+\nu+\lambda+1 \\ 2s+\nu \end{matrix}\right) = \begin{cases} \dfrac{2^\lambda}{\Gamma(\lambda+1)} z^{s+\frac{\nu}{2}}(1-\sqrt{z})^\lambda, & (|z|<1) \\ 0, & (|z|>1) \end{cases}$$

(by the formula in [Erde]) and

$$G_{2,6}^{4,0}\left(z \,\middle|\, \begin{matrix} -s-\frac{\nu}{2}+\frac{1}{2}, -s-\frac{\nu}{2}-\lambda \\ ** \end{matrix}\right) = G_{2,6}^{4,0}\left(z \,\middle|\, \begin{matrix} -s-\frac{\nu}{2}+1, -s-\frac{\nu}{2}-\lambda \\ \dagger \end{matrix}\right),$$

where ** indicates $-s + \frac{\nu}{2} + \frac{1}{2}, -s - \frac{\nu}{2} + \frac{1}{2}, -s - \frac{\nu}{2} - \frac{\lambda}{2} + \frac{1}{2}, -s - \frac{\nu}{2} - \frac{\lambda}{2}, -s + \frac{\nu}{2} + 1, -s - \frac{\nu}{2} - \lambda$ and †, $-s + \frac{\nu}{2} + \frac{1}{2}, -s + \frac{\nu}{2} + 1, -s - \frac{\nu}{2} - \frac{\lambda}{2}, -s - \frac{\nu}{2} - \frac{\lambda}{2} + \frac{1}{2}, -s + \frac{\nu}{2} + 1, -s - \frac{\nu}{2} - \lambda$. Hence it reduces further to

$$
z^{-s-\frac{\lambda}{2}+\frac{1}{4}} \left\{ \frac{2}{\pi} \cos((\nu + \lambda + 1)\pi) K_{2\nu+\lambda+1}(4\sqrt[4]{z}) \right.
$$
$$
\left. + \cos((\nu + 1)\pi) Y_{2\nu+\lambda+1}(4\sqrt[4]{z}) + \sin((\nu + 1)\pi) J_{2\nu+\lambda+1}(4\sqrt[4]{z}) \right\}
$$
$$
= -z^{-s-\frac{\lambda}{2}+\frac{1}{4}} \left\{ (-1)^\lambda \frac{2}{\pi} \cos(\nu\pi) K_{2\nu+\lambda+1}(4\sqrt[4]{z}) \right.
$$
$$
\left. + \cos(\nu\pi) Y_{2\nu+\lambda+1}(4\sqrt[4]{z}) + \sin(\nu\pi) J_{2\nu+\lambda+1}(4\sqrt[4]{z}) \right\}
$$
$$
= z^{-s-\frac{\lambda}{2}+\frac{1}{4}} G^\lambda_{2\nu+\lambda+1}(4\sqrt[4]{z}),
$$

say, where, slightly more general than Wilton's (1.22) [Wilt], we put

$$
G^\lambda_\nu(z) = -(-1)^\lambda \frac{2}{\pi} \sin\left(\frac{\nu - \lambda}{2}\pi\right) K_\nu(z) - \sin\left(\frac{\nu - \lambda}{2}\pi\right) Y_\nu(z) \quad (6.113)
$$
$$
+ \cos\left(\frac{\nu - \lambda}{2}\pi\right) J_\nu(z).
$$

Hence (6.112) reads

$$
\frac{z^{s+\frac{\nu}{2}} 2^\lambda}{\Gamma(\lambda+1)} \sum_{\lambda_k < \frac{1}{z}} \alpha_k \lambda_k^{\frac{\nu}{2}} \left(1 - \sqrt{z\lambda_k}\right)^\lambda \quad (6.114)
$$
$$
= z^{s+\frac{\lambda}{4}-\frac{1}{4}} \sum_{k=1}^\infty \frac{\beta_k}{\mu_k^{\frac{\lambda}{4}+\frac{1}{4}}} G^\lambda_{2\nu+\lambda+1}\left(4\sqrt[4]{\frac{\mu_k}{z}}\right)
$$
$$
+ \sum_{k=1}^L \mathrm{Res}\left(\Delta(w - s)\chi(w) z^{s-w}, w = s_k\right),
$$

which gives a more general form of Wilton's Theorem 1 [Wilt] first given by Tsukada [Tsuk] (cf. [MRSK]).

Rewriting (6.114) slightly, we deduce an analogue of Chandrasekharan and Narasimhan result ([ChNa, Theorem 7.1 (a)]),

Theorem 6.4. *For $x > 0$, the functional equation (6.103) implies the identity*

$$
\frac{1}{\Gamma(\lambda+1)} \sum_{\lambda_k < x} \alpha_k \lambda_k^\nu (x - \lambda_k)^\lambda \quad (6.115)
$$
$$
= x^{\frac{\lambda}{2}+\nu+\frac{1}{2}} 2^{-\lambda} \sum_{k=1}^\infty \frac{\beta_k}{\mu_k^{\frac{\lambda}{2}+\frac{1}{2}}} G^\lambda_{2\nu+\lambda+1}(4\sqrt{\mu_k x}) + \mathrm{P}_\lambda(x),
$$

where

$$P_\lambda(x) = x^{2s+\lambda+\nu} \sum_{k=1}^{L} \text{Res}\left(\Delta(w-s)\chi(w)x^{2(w-s)}, w = s_k\right) \qquad (6.116)$$

and where

$$\Delta(w) \qquad\qquad\qquad\qquad\qquad\qquad\qquad\qquad\qquad (6.117)$$

$$= \frac{\Gamma\left(1-s-\frac{\lambda}{2}-\frac{\nu}{2}-w\right)\Gamma\left(-s-\frac{\lambda}{2}-\frac{\nu}{2}-w\right)}{\Gamma\left(s-\frac{\nu}{2}+w\right)\Gamma\left(s+\lambda+\frac{\nu}{2}+1+w\right)\Gamma\left(1-s-\frac{\nu}{2}-\frac{1}{2}-w\right)\Gamma\left(-s-\lambda-\frac{\nu}{2}-w\right)},$$

with $G^\lambda_{2\nu+\lambda+1}$ *being given by* (6.113).

Corollary 6.1. *For* $x > 0$, *the functional equation* (6.103) *implies the identity*

$$A^\lambda(x) := \frac{1}{\Gamma(\lambda+1)} \sum_{\lambda_k < x} \alpha_k (x-\lambda_k)^\lambda \qquad (6.118)$$

$$= -2^{-\lambda} \sum_{k=1}^{\infty} \beta_k \left(\frac{x}{\mu_k}\right)^{\frac{\lambda}{2}+\frac{1}{2}} F_{\lambda+1}\left(4\sqrt{\mu_k x}\right) + P_\lambda(x),$$

where

$$F_{\lambda+1}(z) = -G^\lambda_{\lambda+1}(z) = Y_{\lambda+1}(z) + (-1)^\lambda \frac{2}{\pi} K_{\lambda+1}(z). \qquad (6.119)$$

We are now in a position to prove an analogue of [ChNa, Theorem 7.1 (b)] (although Theorem 6.4 contains [ChNa, Theorem 7.2], too) by the Riemann-Liouville fractional integral transform.

Lemma 6.1. (Riemann-Liouville integral of Bessel functions) *For the well-known Bessel functions* J *and* Y, *we have*

$$\frac{1}{\Gamma(\mu)} \int_0^y (y-x)^{\mu-1} x^{\frac{1}{2}\nu} J_\nu(ax^{\frac{1}{2}}) \, dx = 2^\mu a^{-\mu} y^{\frac{1}{2}\mu+\frac{1}{2}\nu} J_{\mu+\nu}(ay^{\frac{1}{2}}) \quad (6.120)$$

$(\text{Re}\,\mu > 0, \text{Re}\,\nu > 0).$

$$\frac{1}{\Gamma(\mu)} \int_0^y (y-x)^{\mu-1} x^{\frac{1}{2}\nu} Y_\nu(ax^{\frac{1}{2}}) \, dx \qquad\qquad (6.121)$$

$$= 2^\mu a^{-\mu} y^{\frac{1}{2}\mu+\frac{1}{2}\nu} J_{\mu+\nu}(ay^{\frac{1}{2}}) + \frac{\Gamma(\nu+1)}{\Gamma(\mu)\pi} 2^{\nu+2} a^{-\mu} y^{\frac{1}{2}\mu+\frac{1}{2}\nu} S_{\mu-\nu-1,\mu+\nu}(ay^{\frac{1}{2}})$$

$(\text{Re}\,\mu > 0, \text{Re}\,\nu > -1),$

where S *stands for the Lommel function.*

(6.120) is [Erd1, (63),p. 194] and (6.121) is [Erd1, p. 196]. We only need (6.121) and (6.121) is for treating the J-Bessel function.

Arguing in the same way as in [ChNa], we may prove

Theorem 6.5. *With a C^∞-function R_ρ and a certain constant c we have*

$$A^\rho(x) - R_\rho(x) = c \sum_{k=1}^{\infty} \beta_k \left(\frac{x}{\mu_k} \right)^{\frac{\rho}{2}+\frac{1}{2}} Y_{\rho+1} \left(4\sqrt{\mu_k x} \right), \qquad (6.122)$$

for integral λ, $\rho = \lambda + \alpha$, $0 < \alpha < 1$, $\lambda \geq 2\sigma_\psi - \frac{3}{2}$.

Bibliography

[Apo1] T. M. Apostol, On the Lerch zeta function, *Pacific J. Math.* **1** (1951), 161-167.

[Apo2] T. M. Apostol, Addendum to 'On the Lerch zeta function', *Pacific J. Math.* **2** (1952), 10.

[Apo3] T. M. Apostol, Remark on the Hurwitz zeta function, *Proc. Amer. Math. Soc.* **2** (1951), 690–693.

[Apo4] T. M. Apostol, *Mathematical analysis*, Addison-Wesley, Reading 1957.

[Ausl] L. Auslander, R. Tolimieri and S. Winograd, *Hecke's theorem in quadratic reciprocity*, finite nilpotent groups and the Cooley-Tukey algorithm, Adv. Math. **43** (1982), 122-172.

[Bell] R. Bellman, *Wigert's approximate functional equation and the Riemann zeta-function,* Duke Math. J. **16** (1949), 547-552.

[Ber1] B.C. Berndt, *Character analogues of the Poisson and Euler-Maclaurin summation formula with applications,* J. Number Theory **7** (1975), 413-445.

[Ber2] B. C. Berndt, *Classical theorems on quadratic residues,* Enseign. Math. (2) **22**, (1976),261-304.

[BKTS] R. Balasubramanian, S. Kanemitsu and H. Tsukada, *Contributions to the theory of the Lerch zeta-function,* Proc. Intern. Conf.–The Riemann zeta-function and related themes (2006), 29-38.

[Böhh] P. E. Böhmer, *Differenzengleichungen und Bestimmte Integrale,* Koecher Verlag, Berlin, 1939.

[BShh] Z. Borevič and I. Šafarevič, *The theory of numbers,* Izd. Nauka, Moscow 1964; German transl. *Zahlentheorie,* Birkhäuser, Basel and Stuttgart, 1966; English transl. Academic Press, New York etc., 1966.

[CaoC] J.-Y. Cao and B.-G. Cao, *Design of fractional order controller based on particle swarm optimization,* Intern. J. Control, Automation, and Systems, **4** (2006), 775-781.

[ChNa] K. Chandrasekharan and R. Narasimhan, *Functional equations with multiple gamma factors and the average order of arithmetical functions,* Ann. Math. (2) **76** (1962), 93-136.

[Comt] L. Comtet, *Advanced combinatorics: The art of finite and infinite expan-*

sions, D. Reidel Publ. Co. Dordrecht-Boston, 1974.

[Dave] H, Davenport, *Multiplicative Number Theory*, Markham 1967, second edition Springer 1982.

[Dick] Dickson, *History of number theory*, Chelsea New York 1952.

[DrSh] R. E. Dressler and E. E. Shult,*A simple proof of the Zolotareff-Forbenius theorem*, Proc. Amer. Math. Soc. 54 (1975), 53-54.

[DSSi] K. Diclher, L. Skula and I. Sh. Slavutsukii, *Bernoulli Numbers Bibliography (1713-1990)*. Queen's UP. Kingston, 1991.

[Erde] A. Erdélyi, W. Magnus, F. Oberhettinger and F. G. Tricomi, *Higher transcendental functions*, Vols 1-3, McGraw-Hill. New York 1953.

[Erd1] A. Erdélyi, W. Magnus, F. Oberhettinger and F. G. Tricomi, *Tables of integral transforms*, Vols 1-3, McGraw-Hill. New York 1953.

[Eule] L. Euler,*Remarques sur un beau rapport entre les séries des puissances tant directes que réciproques*, Mem. Acad. Sci. Berlin [17] (1761), 1768, 83-106, Lu en 1749 (Opera Omnia I-15, 70-90).

[Ferg] R. Ferguson, *An application of Stieltjes integration to the power series coefficents of the Riemann zeta function*, Amer. Math. Monthly, **70** (1963), 60-61.

[Frob] G. Frobenius, *Über die Bernoulli'shen Zahlen und die Euler'schen Polynome*, Sitzungsber. Akad. Wissensch. Berlin **2** (1910), 809-847 (p. 826)=*Ges. Abh.* 440-478 (p. 457).

[Funa] T. Funakura, *On Kronecker's limit formula for Dirichlet series with periodic coefficients*, Acta Arith. **55** (1990), 59–73.

[Galo] I. Gal, *Lectures on number theory*, Jones Letter Service, Minneapolis 1961.

[Gold] L. J. Goldstein, *Analytic number theory*, Prentice-Hall, New Jersey 1971.

[Grod] F. S. Grodins, *Control theory and biological systems*, Columbia Univ. Press, New York and London 1963.

[Hass] H. Hasse, *Vorlesungen über Zahlentheorie*, 2 auf. Springer, Heildelberg 1964.

[Has1] H. Hasse, On a question of S. Chowla, *Acta Arith.* **18** (1971), 275–280.

[Hatt] A. Hattori, *Modern algebra*, Asakura-shoten, Tokyo 1968 (in Japanese).

[Haya] Y. Hayakawa, *Systems and their controlling*, Ohmsha, Tokyo 2008 (in Japanese).

[Hel1] J. W. Helton, The distance of a function to H^∞ in the Poincaré metric; elctrical power transfer, *J. Funct. Anal.* **38** (1980), 273-314.

[Hel2] J. W. Helton, Non-Euclidean funicitional analysis and electronics, *Bull. Amer. Math. Soc.* **7** (1982), 1-64.

[Heck] E. Hecke, *Vorlesungen über die Theorie der algebraischen Zahlen*, Akad. Verlagges., Leipzig 1923; English transl: *Lectures on the theory of algebraic numbers*, Springer, New York-Heildelberg-Berlin 1981.

[HKTo] M. Hashimoto, S. Kanemitsu and M. Toda, *On Gauss' formula for ψ and finite expressions for the L-series at 1*, J. Math. Soc. Japan **60** (2008), 219-236.

[HLeo] H. W. Leopoldt, *Eine Verallgemeinerung der Bernoullischen Zahlen*, Abh. Math. Sem. Univ. Hamburg **22** (1958) 131-140.

[HeMe] J. W. Helton and O. Merino, *Classical control using H^∞ methods, Theory,* *optimization and design*, SIAM, Philadelphia 1998.

[Hard] G. H. Hardy and M. Riesz, *The general theory of Dirichlet series*, Cambridge UP, Cambridge 1915; reprint, Hafner, New York 1972.

[HaWr] G. H. Hardy and E. M. Wright, *An introduction to the theory of numbers*, Oxford UP, Oxford 1932.

[Ivic] A. Ivić, *The Riemann zeta-function, Theory and applications*, John Wiley and Sons, New York 1985; reprint, Dover, New York 2003.

[Iwas] K. Iwasawa, *Lectures on p-adic L-functions*, Ann. Math. Studies 74, Princeton UP, Princeton, 1972.

[JoJm] W. Johnson abd K. J. Mitchel,*Summation of sums of the Legendre symbol*, Pacific J. Math. **69** (1977), 117-124.

[Jori] H. Joris, *On the evaluation of Gaussian sums for non-primitive Dirichlet characters*, Enseign. Math. (2) **23** (1977), 13-18.

[Kan1] S. Kanemitsu, *Class field theory through examples, A mini course at Shandong University*, March and August 2007 (unpublished).

[KaKi] S. Kanemitsu and I. Kiuchi, *Functional equations and asymptotic formulas*, Mem. Fac. Gen. edu. Yamaguchi Univ. **28** (1994), 9-54 (in Japanese).

[Keun] F. Keune,*Quadratic reciprocity and finite fields*, Nieuw Arch. Wisk. 263-266.

[Kimu] H. Kimura, *Chain scattering approach to H^∞-control*, Birkhäuser, Boston-Basel-Berlin 1997.

[Kita] Y. Kitaoka, *Introduction to algebraic number theory*, to appear.

[KKPa] S. Kanemitsu, T. Kuzumaki and F. Pappalardi, *Arithmetical class number formula for certain quadratic fields*, to appear.

[KMZh] S. Kanemitsu, J, Ma and W.-P. Zhang, *On the discrete mean value of the product of two Dirichlet L-functions*, Abh. Math. Sem. Univ. Hamburg, **79** (2009), 149-164.

[KTYZ] S. Kanemitsu, Y. Tanigawa, M. Yoshimoto and W. -P. Zhang, On discrete mean value of L(1,), **Math. Z.** 248 (2004), 21-44.

[KTZh] S. Kanemitsu, Y. Tanigawa, and J.-H. Zhang, *Evaluation of Spannenintegrals of the product of two zeta-functions*, Integral transforms and special functions **19** (2008), 115-128.

[Kuro] A. G. Kurosh, *The theory of groups*, Vol. II, AMS-Chelsea, Providence R.I. 1979.

[KUWa] S. Kanemitsu, J. Urbanowicz and N. -L. Wang, *On some generalizations of congruences obtained by Z.-H. Sun*, to appear Acta Arith.

[KUW1] S. Kanemitsu, J. Urbanowicz and N. -L. Wang, *On some further generalizations of congruences obtained by Z.-H. Sun*, to appear.

[KaMa] M. Katsurada and K. Matsumoto,*The mean values of Dirichlet L-function at integer points and class numbers of cyclotomic fields*, Nagoya Math. J. **134** (1994), 151-172.

[Knop] M. I. Knopp, *Hamburger's theorem on $\zeta(s)$ and the abundance principle for Dirichlet series with functional equation*, Number Theory ed. by R. P. Bambah et al, Hindustan Books Agency, 2000, 201–216.

[Kosh] N, S. Koshlyakov,*Investigation of some questions of analytic theory of the rational and quadratic fields, I-III (Russian)*, Izv. Akad. Nauk SSSR, Ser.

Mat. **18** (1954), 113–144, 213–260, 307–326; Errata **19** (1955), 271.

[KuLa] D. S. Kubert and S. Lang, *Modular units*, Springer Verl. Berlin-Heidelberg etc. 1981.

[Lou1] S. Louboutin, *Quelques formules exactes pour des moyennes de functions L de Dirichlet*, Canad. Math. Bull. **36** (1993), 190-196; Correction, ibid. 37(1994), 89.

[Lou2] S. Louboutin, *On the mean value of* $|L(1,\chi)|^2$ *for odd primitive Dirichlet characters*, Proc. Japan Acad. Ser. A **75** (1999), 143-145.

[Lou3] S. Louboutin, *The mean value of* $|L(k,\chi)|^2$ *at positive rational integers* $k \geq 1$, Colloq. Math. **90** (2001), 69–76.

[Lang] S. Lang, *Algebraic number theory*, Addison-Wesley, New Yrok 1970; 2nd ed., Springer Verl., New York etc. 1994.

[Lehm] D. H. Lehmer, *Euler constants for arithmetic progressions*, *Acta Arith.* **27** (1975), 125–142; *Selected Papers of D. H. Lehmer*, Vol. **II**, 591–608, Charles Babbage Res. Center, Manitoba, 1981.

[Lemm] F. Lemmermeyer, *Reciprocity laws: from Euler to Eisenstein*, Springer Verl., Berlin-Heidelberg etc. 2000.

[CPLu] C.-P. Lu, *On the unique factorization theorem in the ring of number theoretic functions*, Illinois J. Math. **9** (1965), 40-46.

[LiuZ] Y.-N. Liu and W.-P. Zhang, *On the mean value of* $L(m,\chi)L(n,\bar{\chi})$ *at positive integers* $m, n \geq 1$, Acta Arith. **122** (2006), 51-56.

[Mats] K. Matsumoto, *Recent developments in the mean square theory of the Riemann zeta and other zeta-functions*, Number Theory, ed. by R. P. Bambah et al, Hindustan Book Agency, 2000, 241–286.

[MacC] P. J. MacCarthy, *Introduction to arithmetical functions*, Springer Verl. New York-Berlin-Heidelberg-Tokyo 1986.

[Mozz] C. J. Mozzochi, *A simple proof of the Chinese remainder theorem*, Amer. Math. Monthly **74** (1967), 998.

[MuPa] M. Ram Murty and A. Pacelli, *Quadratic reciprocity via theta functions*, Proc. Intern. Conf.–Number Theory, Ramanujan Math. Soc. Publ. No 1, 2004, 107-116.

[MRSK] S. Kanemitsu and H. Tsukada, *Contributons to the theory of zeta-functions: The modular relation supremacy*, World Scientific, Singapore etc. 2012.

[Nie1] N. Nielsen, *Traité élémentaire des nombres de Bernoulli*, Gauther -Villars, Paris, 1923.

[Nie2] N. Nielsen, *Die Gammafunktionen*, Chelsea, New York 1965.

[Nark] W. Narkiewicz, *Elementary and analytic theory of algebraic numbers*, PWN Warszawa 1974; 2nd ed., PWN-Springer 1990.

[Neuk] J. Neukirch, *The Beilinson conjecture for algebraic number fields*, in " *Beilinson's conjectures on special values of L-functions*," ed. by M. Rapoport et al, Academic Press, Boston etc. 1988, 193-247.

[PCho] P. Chowla, *On the class-number of real quadratic fields*, J. Reine Angew. Math. **230** (1968), 51-60.

[PErd] P. Erdös, *On the distribution function of additive functions*, Ann. of Math. (2) **47** (1946), 1-20.

[Prac] K. Prachar, *Primzhalverteilung*, Springer Verlag, Berlin-Heidelberg-New York 1957, 1978.

[Rade] H. Rademacher, *Topics in analytic number theory*, Springer, Berlin 1973.

[RaGr] H. Rademacher and E. Grosswald, *Dedekind sums*, MAA, New York 1972.

[Riel] B. Riemann, *Ges. Math. Werke*, Dover, New York 1978.

[Riol] J. Riordan, *An introduction to combinatorial analysis*, Princeton UP, New Jersey 1978.

[Roma] N. P. Romanoff, *Hilbert spaces and number theory I, II*, Izv. Akad. Nauk SSSR **10** (1946), 3-24; **15** (1951), 131-152.

[SUWa] A. Schinzel, J. Urbanowicz and P. van Wamelen, *Class numbers and short sums of Kronecker symbols*, J. Number Theory **78** (1999), 62-84.

[Serr] J.-P. Serre, *A course in arithmetic*, Springer Verl. New York-Heidelberg-Berlin 1973.

[Sieg] C. L. Siegel, *Symplectic geometry*, Academic Press, New York 1964=*Amer. J. Math.* **65** (1943), 1-86; *Ges. Abh.* II, Springer, Berlin-Heidelberg-New York 1966, 274-359.

[Siva] R. Sivaramakrishnan, *Classical theory of arithmetic functions*, Dekker, New York 1988.

[Smit] R. A. Smith, *A note on Dirichlet's theorem*, Canad. Math. Bull. **24** (1981),379-380.

[SrCh] H. M. S. Srivatava and J. -S. Choi, *Series associated with the zeta and related functions*, Kluwer Academic Publishers, Dordrecht-Boston-London, 2001.

[SUZa] J. Szmidt, J. Urbanowicz and D. Zagier, *Congruences among generalized Bernoulli numbers*, Acta Arith. **3** (1995), 273-278.

[Titc] E. C. Titchmarsh, *The Theory of the Riemann Zeta-Function*, Oxford University Press, 1951; revised version by D. R. Heath-Brown, Oxford University Press, 1986.

[TKub] T. Kubota, *On automorhic functions and the reciprocity law in an algebraic number field*, Lect. Notes in Math., Kyoto Univ. **2**, Kinokuniya, Tokyo 1969.

[Tsuk] H. Tsukada, *A general modular relation in analytic number theory*, in *Number theory: Sailing on the sea of number theory*, World Scientific, 214–236 (2007).

[Tul1] J. P. Tull, *The multiplication problem for Dirichlet series*, Proc. Amer. Math. Soc. **9** (1958), 332-334.

[Tul2] J. P. Tull, *Dirichlet multiplication for lattice point problems*, Duke Math. J. **26** (1959), 73-80.

[Tul3] J. P. Tull, *Dirichlet multiplication for lattice point problems II*, Pacific J. Math. **9** (1959), 609-615.

[Yul4] J. P. Tull, *Average order of arithmetic functions*, Illinois J. Math. **5** (1961), 175-181.

[Vand] H. S. Vandiver, *An arithmetical theory of the Bernoulli numbers*, Trans. Amer. Math. Soc. **51** (1942), 502-531.

[Vist] S. Kanemitsu and H. Tsukada, *Vistas of special functions*, World Scientific, Singapore-London-New York. 2007.

[Vis2] K. Chakraborty, S. Kanemitsu and H. Tsukada, *Vistas of special functions II*, World Scientific, London-Singapore-New Jersey, 2009.

[Waid] C. Waid, *On Dirichlets theorem and infinite primes*, Proc. Amer. Math. Soc. **44** (1974), 9-11.

[Widd] D. V. Widder, *Advanced calculus*, 2nd ed. Dover, New York 1989.

[Wash] L. C. Washington, *Introduction to cyclotomic fields,* Springer Verl. New York, 1982.

[Wilt] J. R. Wilton, *An extended form of Dirichlet's divisor problem*, Proc. London Math. Soc. (2) **36** (1934), 391-426.

[Weil] A. Weil, *Basic number theory*, Springer Berlin etc., 1967.

[Wint] A. Wintner, *Diophantine approximation and Hibert's space*, Amer. J. Math. 66 (1944), 564-578.

[Yama] Y. Yamamoto, *Dirichlet series with periodic coefficients*, Proc. Intern. Sympos. " Algebraic Number Theory", Kyoto 1976, 275-289. JSPS, Tokyo 1977.

[YBCa] M. Ye, Z.-F.Bai and B.-G.Cao, *Robust H2/H infinity control for regenerative braking of electric vehicles,* ICCA 2007. IEEE International Conference on Control and Automation, 2007, 1366-1370.

[ZhXu] W. -P. Zhang and Z.-F. Xu, *On a conjecture of the Euler numbers,* J. Number Theory, **127** (2007), 283-291.

Index

194 *Number theory and its applications*